GIFTS FROM SPACE

L. B. Taylor, Jr.

GIFTS FROM SPACE

How Space Technology Is Improving Life On Earth

Illustrated with photographs

The John Day Company New York

Photographs courtesy of the
National Aeronautics and Space Administration

Published simultaneously in Canada by Fitzhenry & Whiteside Limited, Toronto.
Designed by Patricia Parcell
Manufactured in the United States of America

Library of Congress Cataloging in Publication Data

Taylor, L B Gifts from space.
Bibliography: p. Includes index.
SUMMARY: Surveys the benefits derived through everyday use of space technology including medical advances, satellite communication, and weather predictions.
1. Technology transfer—Juv. lit.
2. Astronautics—Juv. lit. [1. Technology transfer. 2. Astronautics] I. Title.
T174.3.T35 338.4'562'940973 77-3261
ISBN 0-381-90056-8

10 9 8 7 6 5 4 3 2

This book is for my son, Chris.

I appreciate the special assistance and advice provided me in the preparation of this book by many members of NASA's Public Information Offices. Especially helpful were two dedicated career specialists in Washington, Ken Senstad and Margaret Ware.

CONTENTS

1	Dividends from Space	1
2	Medical Miracles	4
3	Safety Spin-offs	22
4	A Better Cup of Coffee	31
5	Innovations in Industry	37
6	Transportation Transplants	44
7	New Ideas in Energy	53
8	One-World Communications	61
9	Satellite Weather Watch	74
10	Orbital Prospectors	82
11	Environmental Guardians	98
12	Technology Clearinghouse	102
13	A Better Tomorrow	106
	For Further Reading	119
	Index	123

GIFTS FROM SPACE

1 DIVIDENDS FROM SPACE

Within your lifetime, American astronauts have explored the moon, hundreds of satellites have been launched into orbits above the earth, and electronic spacecraft have been rocketed to other planets in our solar system such as Venus and Mars, Mercury and Jupiter.

The Age of Space is still that young. In fact, the first artificial satellite—Russia's Sputnik I—was orbited on October 4, 1957, followed nearly four months later by the first American satellite, Explorer I. The technology breakthroughs and historic milestones continued in rapid succession through the 1960s, leading to the successful exploration of the lunar surface in July 1969 by Americans Neil Armstrong and Edwin Aldrin.

The early years of the space program in the United States were exciting and dramatic, capturing the imagination of the American public and of people in nations all over the world. Men were walking on the face of the moon, a quarter of a million miles away—a feat that had been dreamed about as long as humans had been on earth. It was a time of high adventure and of national pride.

But as the 1970s arrived, and several more teams of astronauts were blasted into space and camped out on the moon, some of the interest and glamour began to wear off. Scientists were still pleased at the new discoveries being made. But the average person began to shrug and say, "So what, we've been there before." Besides, it was terribly expensive to launch men and machines into space, or to send complex electronic packages millions of miles to other planets.

And there were many more immediate needs to be attended to on earth. Millions of people were starving, or were underfed. Millions of people lived in slums. The world's seas and streams and cities were being choked with pollution. It was time to do something about cleaning up the environment, and it would cost a lot of money. Also, the world's natural resources were running out. There were shortages of oil and gas and other energy sources.

So the average American taxpayer began asking: Why should so much money be spent on space programs when there is so much to be done here on earth?

It was a fair enough question.

Fortunately, there was a good answer. Through the first dozen or so years of space flight, the emphasis was on exploration. We were going into the unknown. It was an era of discovery. New problems had to be overcome, such as how to launch rockets through earth's strong pull of gravity, and how to hurl satellites into orbit above our planet. Hundreds of complex technical questions had to be answered. Could humans survive in space? If so, how could they safely be flown to and from space? To and from the moon?

It took tens of thousands of scientists, engineers, technicians, and skilled craftsmen, and billions of dollars to solve most of the purely physical problems of conquering space. Once that was done, once we had proven we could send astronauts to other celestial bodies and return them safely home, the first major phase of space exploration had been accomplished.

It was time, as the 1970s arrived, to redirect the primary thrust of the national space program from one of exploration to one of exploitation. That is, it was time to begin using the knowledge and technology gained through the early years of the space program for the betterment of life on earth. This became one of the main objectives of the National Aeronautics and Space Administration (NASA).

Today, the results of this space exploitation—the benefits being derived through everyday use of space technology—affect our lives in a great number of ways. Examples are all around us, although we don't always recognize them as such, because we already take many of them for granted.

Just what are the spin-off benefits from space, and how are they affecting and improving our lives?

2 MEDICAL MIRACLES

In one of the finest, most modern medical facilities in the world—Stanford University Hospital in Palo Alto, California—twenty-five-year-old Mrs. Mary Phillips was bleeding to death, and a team of doctors had run out of conventional ways to save her. Over a five-week period, they had given her forty-six pints of whole blood and sixty-four units of plasma, and had performed nine operations. But they could not halt the persistent abdominal bleeding.

Dr. H. Ward Trueblood, chief resident in surgery at the hospital, then called scientists at NASA's Ames Research Center in nearby Mountain View, explained the situation, and asked for help. A team of specialists studied the unusual problem and came up with an unprecedented solution. They recommended that Mrs.

Phillips be fitted in an astronaut-type pressure suit, worn by test pilots to avoid blacking out during high-speed aerial maneuvers. It applies pressure to counter the draining of blood from the brain and upper body.

The suit stopped Mrs. Phillips's internal hemorrhaging overnight by reducing the difference in pressure between the blood within her arteries and the tissue outside them. This allowed her blood to coagulate naturally, and she soon returned home and resumed a normal life.

A few months later, a medical aide in the remote northern Alaskan village of Allakaket, which has a population of only 125, needed emergency help for a seriously ill eleven-year-old girl named Sally Sam.

Allakaket, like many towns in Alaska, had been set up with ground station antennas capable of bouncing messages off a NASA Applications Technology Satellite (ATS). The medical aide tried desperately by this means to contact the U.S. Public Health Service Hospital in the city of Tanana, on the Yukon River one hundred miles south.

For some reason the aide couldn't rouse Tanana at that time. But he did get through to an ATS control station in Mojave, California, which, in turn, relayed the message by satellite 22,000 miles above earth to Anchorage, Alaska, and from there up to Tanana—all in a matter of minutes. The Public Health Service physicians there diagnosed the illness as acute appendicitis. Fifteen minutes later an evacuation aircraft, with doctor aboard, took off from Tanana to Allakaket. The little girl was taken to the hospital for surgery and recovered fully.

Mrs. Phillips and young Sally Sam are both alive today as a result of technological spin-off benefits from the U.S. space program. The advanced processes used to save them are but two of thousands of research discoveries and engineering innovations—developed as astronauts headed for the moon, and as spacecraft sped to distant planets—that are now being adapted to at-home applications in the science of medicine. To help speed this transition, NASA has assigned specially trained personnel to work directly with medical researchers to channel complex data into everyday, in-the-field use.

With heart-attack victims, every second between the initial strike of pain and the administering of professional help can mean the difference between life and death. More than 60 percent of the fatalities from this disease occur within an hour after the first attack; yet, up to now, the ambulance transit period is often lost time in terms of diagnosis and treatment.

The Space Age is changing that. Through an ingenious system originally designed to check astronauts' heart action, electrodes are "sprayed" on a patient's body in the ambulance. That is, a sticky chemical compound is applied directly to the skin. An adhesive disk is placed on this, and the electrode is attached to the disk. This way, an electrocardiogram (EKG) can be flashed via a radio-telephone link to the hospital while the vehicle is en route. Reading the EKG at a hospital console, the doctor has advance knowledge of the patient's condition and can make whatever preparations are necessary before the patient arrives, saving precious minutes. The system is being used successfully in Los Angeles, Denver, Houston, and other metropolitan areas.

The increasing use of space-related electronic devices and techniques promises substantial reductions in hospital operational expenses, as it frees highly trained personnel from routine patient-watching duties. One case in point is a tiny sensor and radio transmitter, first developed under NASA sponsorship, that has been modified to monitor infants and comatose adults suffering from windpipe obstructions. When injury or disease causes blockage of the upper air passage to the lungs, surgeons frequently must perform a tracheotomy—inserting a small tube in the throat to enhance breathing. If this tube becomes clogged, breathing will stop and brain damage or death can result within two to four minutes.

To prevent this danger, a full-time nurse must constantly check the tube visually and take immediate corrective action if necessary. The new sensor, however, by noting subtle differences in the temperature of air passing through the tube, actuates an audible or visible alarm within ten seconds of any change. The signal can be tied in to a console at a nurse's station, allowing one attendant to keep watch over dozens of patients.

A similar system is now in use on an experimental basis at the Children's Hospital Medical Center in Oakland, California. Here, delicate sensors are attached by microminiature connectors to newborn infants with respiratory ailments. The transmitter sends a pulsating

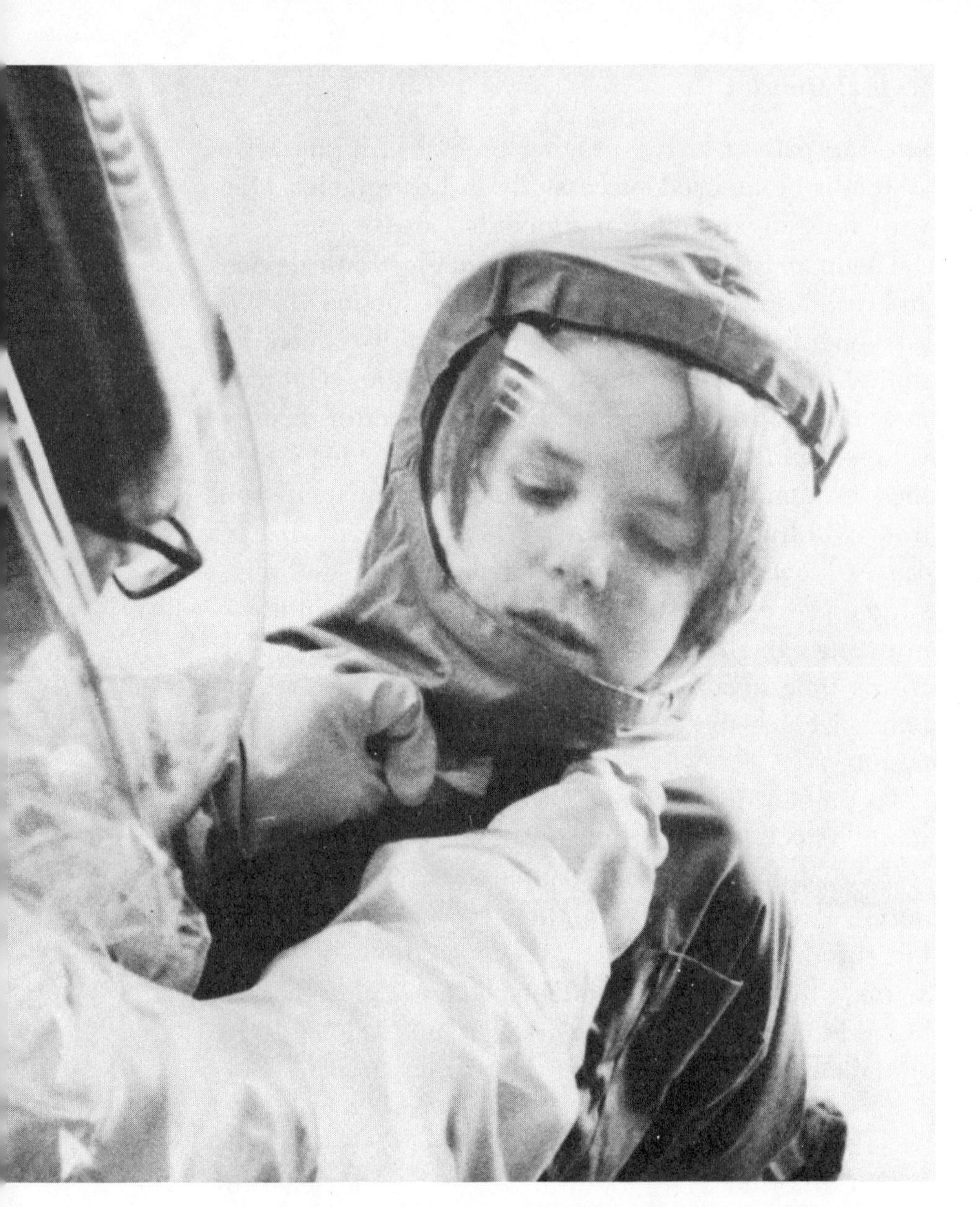

Special protection: biological isolation garments, using astronaut space-suit technology, enable young leukemia patients to move about freely outside sterile treatment rooms.

Looking like a junior space pilot, this young girl can be moved about outside totally protected from germs in a biological isolation garment. It was patterned after suits worn by astronauts on their return to earth from the moon.

signal to a central nurse station. If the baby has trouble breathing, the signal is interrupted, instantaneously warning nearby attendants.

Across the nation in a Miami hospital, small, battery-powered electronic devices manufactured by aerospace companies are strapped to patients' arms and legs in the intensive-care unit. Through these devices, such vital physiological information as temperature and blood pressure can be transmitted from as many as sixty-four patients to a single nurse at a monitoring console.

In other areas around the country all sorts of strange-looking contraptions and paraphernalia—originally designed exclusively for moon-bound astronauts—have been adapted to a number of practical, if unusual, applications for the sick and injured.

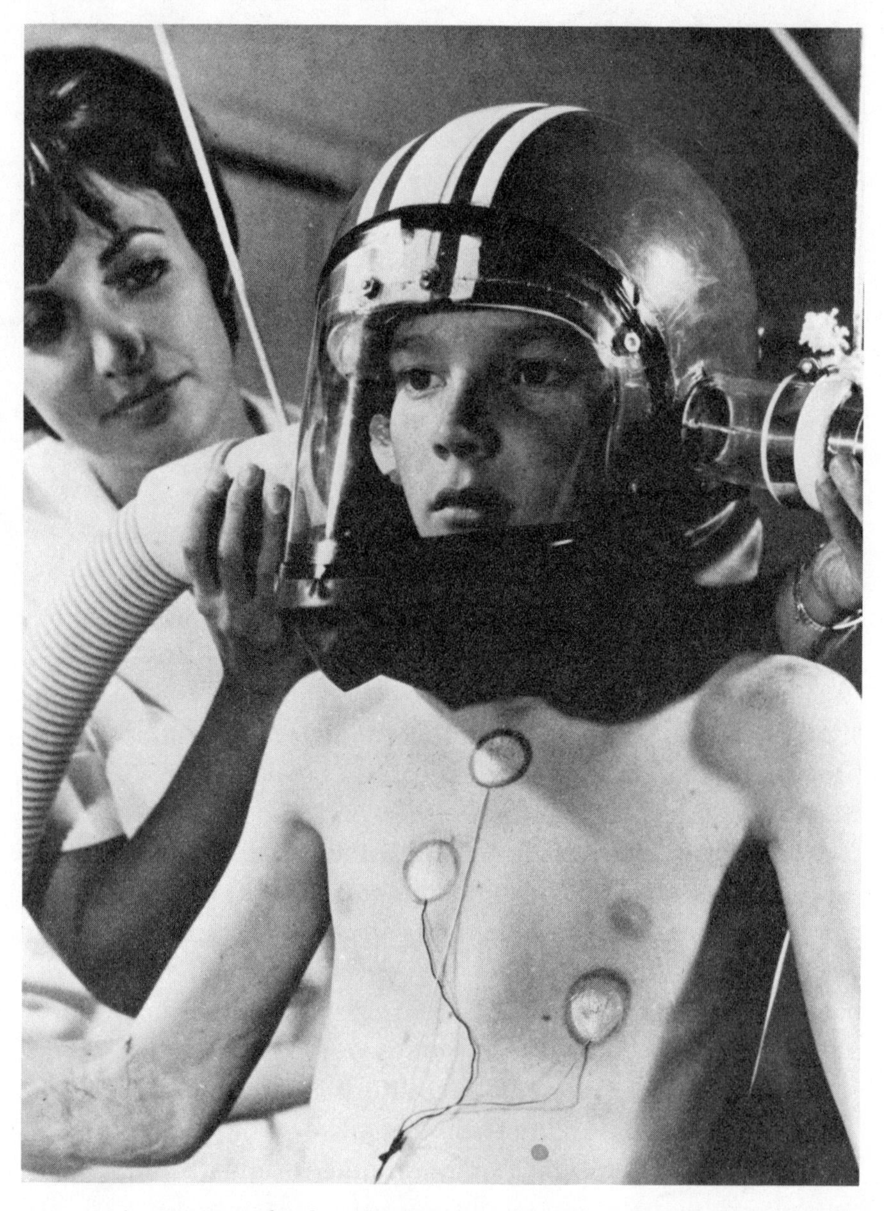

Astronaut-type headgear has found many practical applications in the field of medicine. Here, it is used for a respirometer to measure oxygen consumption while a young patient exercises.

Doctors at the Kansas University Medical Center in Kansas City have found a unique use for astronauts' helmets—as respirometers for children. The conventional rubber mouthpieces formerly used to collect exhaled breath were both uncomfortable and difficult for the youngsters to keep in place. At times this impaired the accuracy of the data on oxygen consumption.

The helmets seem made to order to solve this problem. They come equipped with an air inlet and outlet, a rubber seal around the neck, and a suction pump to circulate fresh air continuously, collect the exhaled breath, and draw the combined fresh air and exhaled breath neatly into an oxygen analyzer. As can be imagined, the kids love to use the helmets. They have turned a troublesome chore into fun.

In some hospitals surgeons, too, are now using astronaut-type helmets—in place of surgical masks. The helmets, in addition to being comfortable, are more sanitary and thus may reduce still further the possibility of infection in the operating room.

Another application for space headgear is being investigated at the Scott-White Clinic and Hospital in Texas. Here, helmets equipped with sponge electrodes to obtain electroencephalographic (EEG) tracings from astronauts and test pilots under stress are being redesigned to detect hearing defects in children.

Even the Apollo spacecraft's specially designed windshields have been put to work on health-care problems. They are used in the treatment of severe burn victims. These curved "shields" are suspended over patients, allowing freer movement while maintaining a constant temperature under their surface. Water loss through evaporation, one of the major problems in the healing process, also is cut down through use of the constant-temperature shield. Doctors have found that in many cases where the shields are applied, pain-suppressant narcotic drugs are not necessary.

Similarly, a dry immersion bed designed for research on metabolic rates of astronauts under the weightless conditions of space now provides relief for victims of severe skin burns or ulcers. In hospitals, patients using the bed lie on a waterproof sheet and remain dry while "floating" in water heated to body temperature. Because the patient is buoyantly supported, there is an infinite number of pressure points, rather than maximum pressure on any single part of the body. The bed is used for people in long-term bed-confinement cases to prevent and to treat pressure-produced skin ulcers and to treat massive skin burns.

Two other developments promise greater convenience and comfort for long-term bedridden patients. Bed sores in particular often bother people who must spend most of their time in a bed or in a sitting position. First developed for seats in astronauts' spacecraft, a new plastic foam may make such people comfortable—in the form of bed pads, cushions for wheelchairs, and as a lining for splints in fracture cases. This padding material can be formed in each case precisely to "fit" each patient's needs.

Even a simple personal chore such as brushing teeth

can be a real problem for those who must stay in bed. Because of problems encountered during weightlessness, astronauts on space missions must swallow their toothpaste after brushing their teeth. Regular toothpaste contains a detergent that makes it foam. NASA has developed a digestible, nontoxic toothpaste that is easily swallowed.

At many of the nation's leading medical and hospital centers a number of additional space-spawned ideas are in various stages of research, experimentation, and laboratory testing. For example, NASA specialists are working with Stanford University School of Medicine scientists on a dramatic new sonar device that will expose secrets about the functioning of the human heart.

Special sonar machines—which emit and receive high-frequency sound waves—can measure precisely the amount of blood pumped out at each contraction of the heart muscle. This is something standard monitoring devices cannot do. The technique can be applied as a screening procedure for patients with known or suspected heart disease, and it can be used to follow progress of the heart's healing process in patients recovering from a cardiac attack or from open-heart surgery.

At the St. Louis School of Nursing and Health Services, a revolutionary "muscle accelerometer" has successfully undergone extensive testing to measure minute muscular tremors in the human body. It was patterned after a "momentum transducer," produced by NASA to measure micrometeorite hits on spacecraft, and is capable of recording impacts as faint as that of a grain of sand dropped from one inch. This may help doctors in early diagnosis of such dreaded neurological ailments as Parkinson's disease.

One of the most intriguing potential applications of space-spawned devices in the medical area involves a new electrostatic camera, which produces moving or still "instant pictures" without any processing. It can focus on a patient in critical condition and can keep vital photographic records instantly available for physicians. Transducer-transmitters that relay intestinal data are currently in use, and doctors now anticipate a battery-powered television system small enough to be swallowed, which would transmit pictures from a patient's stomach!

The list of usable medical spin-offs from space—tangible ones that are helping save lives and mend bodies now, and others that are being converted from research laboratories to practical uses almost daily—includes scores of products, instruments, techniques, and systems. Among them are:

Life-prolonging miniaturized electronic heart pacers.
New alloys for improved artificial limbs.
Electronic devices that convert audio sounds to vibrations that can be "heard" though the sensitive fingertips of the deaf and blind.
Digital computer processes that clarify medical X rays.

Yet all the benefits tapped from space technology to

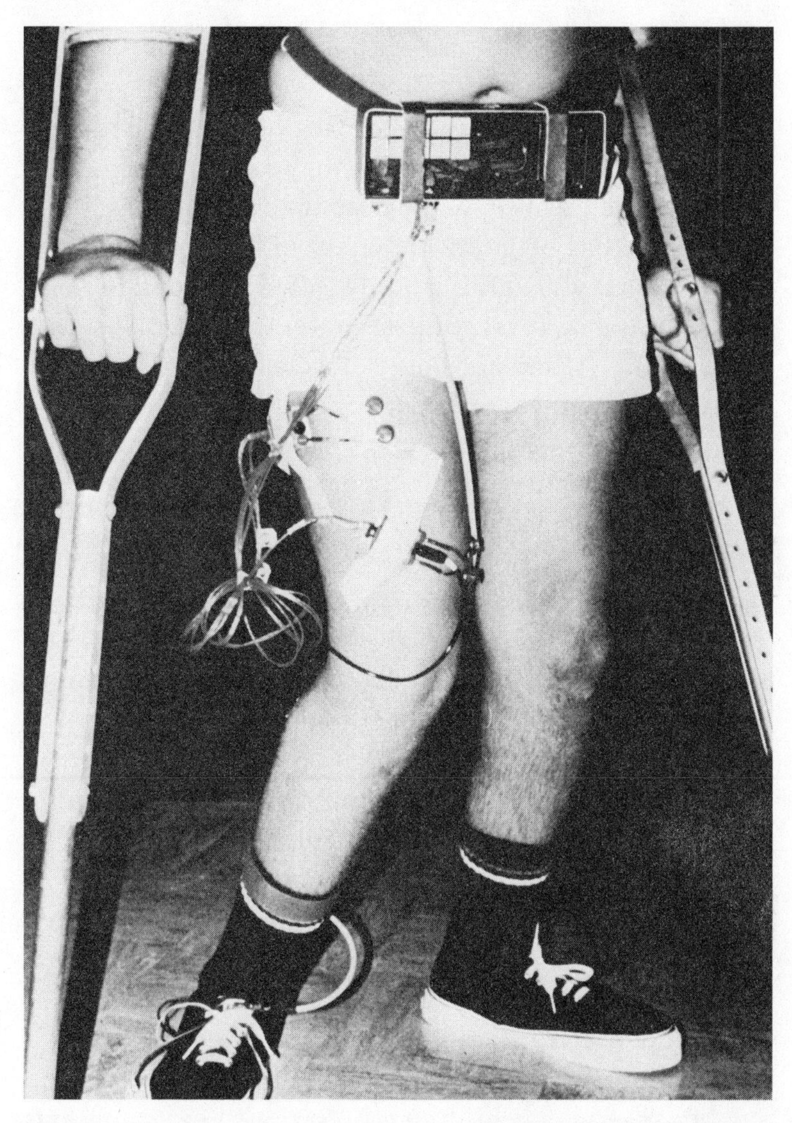

The use of biotelemetry, or radio transmission such as that used to monitor astronauts' bodily functions in space, has found down-to-earth use in medicine. Biotelemetry, for example, monitors the awkward, jerky gait of cerebral-palsied children. It eliminates the previously necessary long bundle of wires leading to recording equipment, making it more convenient for young patients.

date have but scratched the surface of what realistically can be expected in the immediate years ahead.

One of the greatest medical advances in the past quarter century has been development of heart pacemakers. For patients who suffer what is called a complete heart block, pacemakers generate electrical pulses that rhythmically stimulate heart muscles to contract regularly. They help control weak or erratic heart rates. The first pacemakers were attached outside a patient's skin. They were cumbersome, expensive, and their batteries needed replacement frequently.

But space-spawned knowledge has greatly improved pacemakers in recent years. Nickel-cadmium battery cells, developed to improve rechargeable spacecraft batteries, now provide wearers ten to twenty years of reliable use on pacemakers implanted under the skin, outside the rib cage. At first, wearers got only two to five years of such use, before they required an operation for replacement of the pacemaker. The newer units can be recharged at home in just ninety minutes.

Other pacemaker models have been greatly reduced in size—to a unit smaller than a cigarette package—through the use of microelectronic circuitry, which has improved electrical efficiency by 45 percent and reduced interconnections by 50 percent. Because of the smaller size, many children with serious heart problems now can wear pacemakers for the first time.

Over the past several years, NASA has applied its expertise to developing teleoperator and robot technology to help busy astronauts in spacecraft, and to build

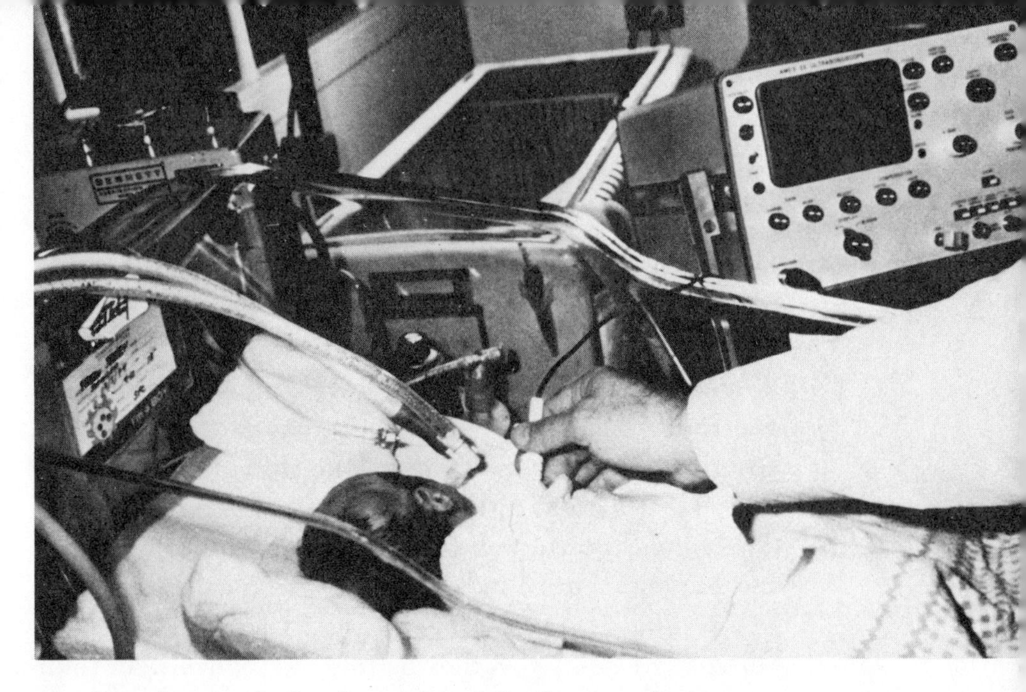

An echo-cardioscope, developed to monitor cardiac functions of astronauts in flight, has been adapted for hospital use. It forms images of internal structures using high-frequency sound in much the same way that submarines detect underwater objects with sonar.

mechanical arms and legs to carry out experiments on the moon or other planets. Much of this skill and knowledge, with a little adaptation, is being used to benefit severely handicapped people on earth.

It is estimated that nearly two million people in the United States, and millions more elsewhere, are afflicted with varying degrees of paralysis. Nearly two-thirds of these individuals have severe deficiencies in manipulative or mobility capabilities.

One of the most extraordinary of these space "adaptations" is a sight switch, first used by astronauts, that has been converted successfully for quadriplegics—persons with no use of their arms and legs. The switch enables a patient to manipulate a motor-driven wheelchair by eye movement. It operates on the principle of infrared reflection from the eyeball. An infrared light source bounces light off the white of the eye into a photoelectric cell, which carries the message to a control activator. When the eyes are moved sideward, for example, one eye reflects the light while the pupil of the other eye absorbs it. This causes an imbalance of voltage that controls the direction of the wheelchair. Patients also use the switch to turn pages of a book, flip a thermostat on and off, and control radio and television sets.

A similar device has been tested successfully and is being used today by other patients without use of their arms or legs. Mrs. Celeste Thompson contracted a severe case of poliomyelitis when she was nineteen years old and was left totally paralyzed, with the exception of some head and neck movement, and a slight motion in her left thumb.

Bioengineers at the Rancho Los Amigos Hospital near Los Angeles worked closely with NASA to develop some new equipment for Mrs. Thompson that is controlled by a sensitive tongue-pressure switch. She triggers a number of complex movement and direction switches by applying slight pressure with her tongue. The first day she used the device, Mrs. Thompson wrote her first letter in eighteen years. Today, she can type twenty-four words per minute and operates a telephone-answering service.

A muscle-oriented device, designed by technicians at NASA's Kennedy Space Center in Florida, provides new mobility for paralyzed human muscles. It does this through a tiny connector that applies small electrical currents from an outside power source through minute terminals embedded in the patient's skin. Long aware that an electrical pulse can make a muscle move, doctors have discovered that tiny platinum wires, attached to small pads on certain nerve endings, can be brought up through the surface of the skin and attached to external power supplies to move these muscles.

The embedded implant is only one-fourth the size of a dime. The portion above the skin surface is a minuscule grooved pin that accepts an open-end, fork-type connector, providing two or more points of secure electrical contact.

A number of space-related advances are being made to aid amputees. The Marshall Space Flight Center in Alabama, for instance, working closely with specialists at a California hospital, has built an artificial hand. It includes a unique finger-control capability that allows amputees to pick up and operate standard power tools.

Marshall scientists have also helped perfect a high-purity, high-strength carbon that is being used for artificial-limb attachment sockets implanted in the bone. The technology was first used for reentry heat shields and rocket-nozzle liners. The new material provides the best available combination of strength, light weight, and physiological compatibility for implanted devices. It combines with skin and bone to form permanent body "plug-in." This direct skeletal attachment improves the wearability of artificial limb devices.

Scientists and engineers in the Composite Materials

Laboratory of NASA's Langley Research Center in Virginia have built a leg brace of lightweight graphite epoxy composite material to replace regular orthopedic leg and pelvic braces. Conventional braces, made of steel, aluminum, leather or wood blocks, and some form of padding, are too heavy and impair patient movement.

In the new design, the strut members and waistband pieces of an ordinary brace were replaced by the composite material, which weighed less than half the original metal components of the same strength. The new brace prototypes, being worn by patients at a number of hospitals and clinics in the southern United States, have proved far more mobile and are more attractive.

Perhaps one of the most unusual space-to-medical spin-offs was first tried at the Texas Institute for Rehabilitation and Research. Here, handicapped children learn to walk in a sling-support outfit invented at NASA's Langley Center to help acquaint moon-bound astronauts with the one-sixth gravity conditions they would encounter on the lunar surface. The rig consists of a walkway, tilted off the horizontal, on which young patients exercise while suspended sideways by belts and pulleys. Various modifications of this sling are being used at other sites across the nation to help bedridden patients to retrain unused muscles to walk again.

A number of important aids to the blind and deaf also are being derived from space research. One corporation that makes instruments for spacecraft is looking into an electronic sight aid for the blind. The company has already adapted the small electronic sensing devices used in spacecraft for another use—restoration of hearing to the deaf by surgical implantation.

Another corporation has successfully accomplished eye surgery with a pinpoint of intense light from a laser, and company specialists believe the laser can also be used in eye-tumor removal and brain surgery.

A NASA technique for semiautomatic inspections of microfilm records has led to the invention of a small machine to help blind persons identify paper money. It is done through a sound "signature." As a bill is passed under a light source, a phototransistor measures changes in its light patterns and an oscillator converts them into sound signals. The sounds are like beeping tones heard on long-distance phone calls. Because each denomination of currency has a different design—George Washington is on the one dollar bill, Abraham Lincoln on the five, etc.—each gives off a different tone under the light source. With a little practice, blind people can easily identify the differences.

These are but a representative few of the hundreds of space-engineered products, techniques, and systems that are being adapted for use by medical experts to enhance the rehabilitation of patients. In many cases, these advanced developments offer new hope for the handicapped.

3 SAFETY SPIN-OFFS

Today millions of people in the United States and in other countries around the world are leading safer lives, thanks to technology benefits spun off from the space program. Home and working environments have been made safer. Automobiles, airplanes, trains, highways, and bridges are safer. The same super-stringent reliability standards used to build spacecraft and rockets have been converted to decrease greatly failures in structures, mechanical devices, and machines. And this has led to safer working conditions in factories and plants, safer performance of machines and equipment in the home, and fewer hazardous conditions in buildings and offices.

Perhaps nowhere has the payoff impact been more pronounced than in the field of fire safety. Ironically, this is due in part to a tragedy. It occurred at the Kennedy Space Center in Florida in January 1967, when three astronauts were burned to death in their spacecraft while still on the ground during a prelaunch test. As a result, NASA launched a massive research program, and today the space agency has one of the most complete data banks ever compiled on the burning characteristics of most materials available for testing. Throughout the research and development work, NASA has maintained a close liaison and testing program with fire-fighting and fire-preventive associations to achieve practical designs for down-to-earth uses.

This has led to new, safer apparel for the nation's firemen. Items include new thermal underwear, coveralls, chaps, trousers, jackets, caps worn under helmets, gloves, nonflammable boots, and "proximity suits" that permit fire fighters to move closer to fires or even enter the flames if necessary. The nonflammable materials used in these garments were first developed to insure maximum safety of moon-bound Apollo crews in the oxygen-rich atmosphere of their spacecraft.

More recently, NASA has perfected a fire fighter's breathing apparatus that is being used in New York City, Los Angeles, Houston, and other major metropolitan areas in the United States. It replaces conventional breathing systems, which many firemen found cumbersome and which restricted their vision. The new units are up to 40 percent lighter, and feature an improved face mask that allows better vision and a unique high-pressure, longer-duration air tank. The air tank is made of spiral-wound glass fiber over an aluminum liner—a

Fire-fighting advancements, encouraged by use of space developed technology, include a new compressed air breathing unit. It is lighter, and increases firemen's mobility by 40 percent.

technique originally developed for lightweight, solid propellant rocket motor cases.

The new unit has been called the first major improvement in compressed-air breathing systems in over twenty years, and has been enthusiastically received by fire-fighting professionals who have used them.

In addition to apparel, NASA has fostered a number of other fire safety improvements. These include several fire-retardant or nonflammable foams, paints, fabrics, and glass fiber laminates. To help avert major disasters, experts at NASA's Ames Research Center are working on fire-resistant structural panels for use in high-rise buildings and low-cost housing such as mobile homes.

Under special study are fire-retardant coatings and wood veneer, used to cover resin-bonded composite wood boards and plastics. At the Johnson Space Center many of these and other materials are being test-applied to fire-resistant carpeting, seats, headrests, paneling, curtains, and fire walls.

New York City had made other use of NASA's fire-fighting expertise. A tragic fire in 1973 at a liquefied–natural-gas storage tank on Staten Island killed forty people. To prevent recurrence of the disaster, New York's Fire Commissioner called on NASA for help. The space agency, which has an impressive safety record in handling highly volatile rocket fuels, has helped the city develop a comprehensive safety program for the storage and handling of liquefied natural gas.

Fire-safety developments are not limited to the land, however. NASA and the U.S. Coast Guard have worked together to design and build a lightweight, portable fire-fighting module for use in battling shipboard and dock fires. The module contains its own pumps, hose, fire-fighting suits, and other equipment. It can be lifted by helicopter or dock crane onto the deck of almost any type of boat, and is capable of pumping water from the sea at a rate of up to 2,000 gallons per minute for up to three hours.

Through these and other programs, space spin-off benefits are helping cut down on the two to three million preventable fires in the United States each year—fires that annually claim from 10,000 to 15,000 lives, and more than $10 billion in property damage.

Other benefits are being adapted to make the nation's roads safer. One of these has evolved through a NASA program designed to prevent hydroplaning, or skidding by aircraft, on rain-drenched airport runways. Through exhaustive research, it was determined that a "grooving" technique prevents formation of a continuous water film on runway surfaces. The groove lines carry off standing water. This insures traction for aircraft tires, permitting pilots to apply brakes without skidding on a "film" of water. National Airport in Washington, D.C., was grooved in 1967, and the results were so successful that other large airports soon had the grooves cut into their runways.

A similar grooving technique was initiated for the nation's major highways. Results have been equally impressive. State safety officers have reported an 80 to 90 percent reduction in damage, injury, and death from skid accidents on roads that have been grooved.

Another major highway safety improvement, inspired by space technology, has been new highway crash barriers, capable of absorbing the impact of an automobile traveling up to 60 miles per hour. The key here is an energy-absorbing component originally designed for astronaut couches on Apollo spacecraft, to help soften the shock of a rocket's blast-off from the launch pad. The system cuts down collision impact by dissipating heat and absorbing energy when an axial force is applied.

Tires are being made safer through use of an ultrasensitive fast-scanning infrared optical device, first used by NASA for testing miniaturized electronic circuits. The device produces a cathode-ray–tube picture of the heat in tires as they spin rapidly on a special testing rig—up to 200 miles per hour for automobile tires, and twice that fast for aircraft tires. The camera provides what is called a "heat picture" of the tire, in which flaws or hot areas appear to the infrared eye as bright spots.

Even visibility of drivers on the road is being improved through space program fallout. When an astronaut's helmet visor fogged up during a space walk in earth orbit, NASA engineers went to work on the problem. They came up with a special antifogging compound. Today it is being used on automobile windshields, for helmet visors worn by motorcyclists and firemen, and for face masks worn by divers. The compound has also proven effective on plastic, aluminum, and other reflective surfaces, such as protective goggles, where it is desirable to maintain a fog-free state.

The vast range of safety spin-offs extends to waterways,

Greater safety on the seas is possible via a lightweight, inflatable, non-tipp-able radar-reflective life raft now being used by the U.S. Coast Guard and sold commercially. It was originally developed by NASA for American astronauts following splashdown in the ocean upon return from space flights.

too. A new lightweight, inflatable, nontippable radar reflective life raft, for example, is being used by the U.S. Coast Guard and is being sold commercially. Its concept emerged from a survival kit designed for astronauts when they splashed down at sea following their return from space flight. The raft's brilliant reflective orange canopy can be spotted by aircraft radars at distances up to fifteen miles, greatly enhancing the survival prospects for those lost at sea.

As many as 40,000 people a year find themselves in distress situations. These include lost explorers and sportsmen, and people on small boats lost or sunk, downed aircraft, and ocean ships in trouble. NASA is working on development of a search and rescue system that would be operated by using a satellite in earth orbit. People in trouble, equipped with an inexpensive pocket-sized radio, could relay an SOS signal to the satellite, which would pass that signal and its location on to a central rescue headquarters from which help could be dispatched.

Some of the most impressive safety improvements resulting from the fruits of space research are being made in the field of aviation. NASA is particularly concerned with applying technology for preventing accidents caused by air turbulence and air-route congestion.

Studies of wake turbulence—that is, air disturbances caused by the forward motion of aircraft—have led to new standards for spacing planes at busy airports during takeoffs and landings. Other research, in cooperation with the Federal Aviation Administration, is leading to the completion of visual and radar alarm devices to warn pilots of impending collisions with other aircraft.

Tests are also being conducted to find and perfect means of reducing airport fog hazards, which annually cost airlines more than $100 million and create passenger inconvenience in delayed or canceled flights.

And, coming in the not-too-distant future will be safer air-traffic control—from space. A recent National Academy of Sciences study of traffic control and navigation over oceans concluded that satellite systems in earth orbit offer the best way of handling the traffic on the overcrowded air and sea lanes expected in the coming decades. Such a system could provide instant, continuous, and automatic plotting of flights and communications with pilots over vast ocean areas. In addition to improved safety, potential benefits of satellite air-traffic control and navigation include precise and frequent position-fixing, greater reliability, and all-weather operation.

Many other innovative safety designs, techniques, systems, and products are being transferred from space to earth use annually. To spread the word on these new developments, NASA has established an Aerospace Safety Research and Data Institute at its Lewis Research Center near Cleveland, Ohio. Here, a computerized information system has been set up to speed the conveyance of solutions to even the most puzzling safety problems.

4 | A BETTER CUP OF COFFEE

It has become a "buzz" phrase of the times . . . "If we can land men on the moon, why can't we do this . . . or how come we can't do that?" Television commercials show people drinking coffee, frowning at it, and exclaiming, "They can send rockets to Mars and men to the moon, why can't they make a cup of coffee I like without caffeine?"

Behind such expressions is an underlying theme. What, the general public wants to know, is space technology doing for *me?* The answer, in down-to-earth, measurable improvements in consumer and household goods and services is—plenty. In and around the home, direct space spin-off benefits are everywhere.

Food shoppers can select a number of items, such as tea, coffee, soup, potatoes, and onions, in a freeze-dehydrated form. Their arrival was hastened by spacemen's needs on moon flights. Compressed/freeze-dried foods make ideal, compact, emergency rations, and are used widely by hunters, campers, and backpackers.

Also at the supermarket, choice cuts of steak, chops, and other meat-counter items are wrapped in the same type of transparent polyester film used initially for NASA's giant balloon satellites, such as Echo. The film is 0.0005 inch thick. In addition, many high-strength aluminum foils used to protect freeze-dried foods and perishables are made of the same material used on early communications satellites.

There are numbers of space-to-home kitchen applications. For example, thick roasts can now be cooked in half the normal time by inserting thermal cooking pins into the meat, cooking it from the inside out. This evolved from the need to meet heat-transfer requirements for space vehicles. Cooking pins use condensible vapor to transport heat. Meat can also be frozen faster by using them, since the pins can work in reverse to "pump" heat out of the meat.

Hard, heat-resistant coatings applied to certain cookware, such as Teflon, are a fallout from the space program. This heat-shield technology is used to protect manned spacecraft reentering the earth's atmosphere through temperatures that reach up to 5,000 degrees Fahrenheit. Similar technology has been applied to thermos jugs. It is so effective that if steaming hot coffee were poured into a tank covered by this material, the coffee would lose less than one degree of temperature in a year!

In other parts of the home, new sealants, first prepared for caulking seams in spacecraft, now waterproof the gaps between shower tiles. And latex paints, a spin-off from the need for ultraviolet radiation protection, are widely used to decorate house walls.

Digital clocks today operate with accuracy not previously possible. They are made with a precision mechanism much like one designed by NASA to position and test spacecraft models in wind-tunnel experiments. Since moon-flight missions required pinpoint, split-second timing of unprecedented accuracy, a new quartz crystal oscillator was developed. This revolutionary new time-keeping base has been converted to practical purposes on earth in a line of consumer clocks and watches. These intricate pieces maintain accuracy to within one minute per year.

More efficient batteries, too, are a direct result of spin-off. The same nickel-cadmium battery technology used in pacemakers for heart patients has found additional application in the form of power sources for golf and baggage carts, portable medical equipment, photographic flash units, toys, and other appliances. These batteries can be recharged one hundred times faster than conventional ones.

Research that led to a highly reliable flashlight switch used by astronauts on all manned space flights, has been adapted for public consumption. Millions of flashlights have been sold with longer life guarantees; their switches will not corrode and cause battery drain.

Electrical wiring in a typical manned spacecraft, if laid

out in a straight line, would stretch for miles. To cut down on the size and increase the efficiency of such wiring, NASA years ago produced flat electrical conductor cables and low-voltage switching circuits. These are now widely used by home builders. As standard 110-volt wiring can be dangerous, it normally is encased in heavy metal cables and hidden inside walls, which sometimes makes it difficult to reach for repair or for extension to new electrical outlets and ceiling fixtures. But with new systems, the wire is flat and thin enough to be concealed under carpeting, or even pasted on walls and covered with paint or wallpaper.

In the field of fashions and fabrics, the creation of a new type of undergarment for astronauts has led to sales of similar items for consumers. The new underwear, made with a specially designed, texturized fabric, provides unusual support without compromising flexibility, no matter how tight the fit. Other space research in fabric materials has led to a super-insulating vacuum-metalized nylon, or taffeta, that is heat reflective yet porous and machine-washable.

Space-inspired fabric progress is benefiting campers and sportsmen and -women, too. Aluminized Mylar is a prime example. Originally designed for super-insulation in space, it is made by laminating a special plastic material with a sheer coating of aluminum 0.0005 inch thick. It now is being used to make popular lightweight blankets, ski parkas, coats, sleeping bags, and life-raft canopies. The blanket can keep cold out by retaining body warmth, or it can serve as an insulator to keep heat out in the summer. A blanket large enough to cover a king-sized bed weighs only a few ounces, and can be folded up and stuffed into a coat pocket.

The list of consumer-product spin-offs literally covers thousands of items. A way to make a better cup of coffee without caffeine may not have been found in the vast data banks of stored aerospace knowledge—but a lot of far more important discoveries have been and are being produced that are making life more convenient on earth today.

Aluminized mylar developed to make certain satellites more reflective, and for space-suit insulation, has a variety of consumer applications, such as sportsmen's blankets and jackets, ski parkas, sleeping bags, and even life-raft canopies. The blanket, for example, weighs only twelve ounces and can be used equally well to keep heat out or to keep available heat in.

For the sports person, this jacket, fashioned from aluminized mylar, weighs only ten ounces, and absorbs warmth from the sun. The material, spun off from mylar used to make satellites more reflective, is bonded to a tear-resistant fabric to allow radar reflection as well as higher visibility under all light conditions.

5 INNOVATIONS IN INDUSTRY

The swing of space technology to earth-borne uses has proved a boon to business and industry as well as to the individual consumer. Some of this spin-off has been covered in earlier chapters, such as the vast array of medical and safety equipment, devices, systems, and programs being marketed by specialized companies.

Two other giant industries directly benefiting from this windfall are the computer business and the field of microelectronics. It is fair to say that the U.S. space program has been instrumental in helping accelerate the boom-growth of the computer age. Vast amounts of complex scientific and engineering data were required for space flights. Advanced methods of computer technology were developed for the design and production of virtually every component of a spacecraft; for precision

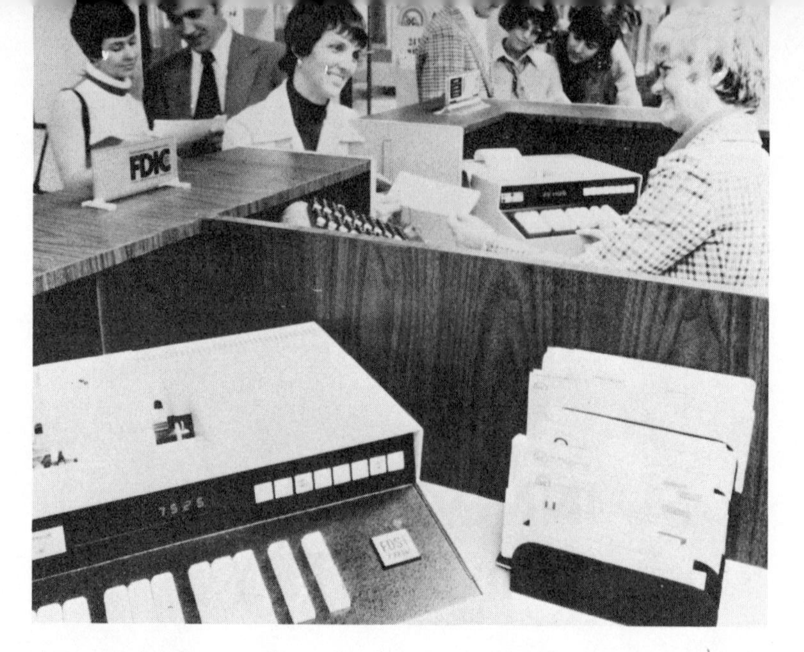

Automatic banking machines that dispense money when you insert a credit card are but one example of space spin-off technology used to improve banking operations. Other applications include automatic credit authorizations and check processing.

control of spacecraft in flight; and for the storage, classification, and retrieval of data from each mission.

The refinement of such special computer "software" has multiple applications in industry. It has, for example, been tailored for use with computers in air-traffic control, industrial-process control, engineering design, automation of hospital services, and for sophisticated medical diagnosis.

Space-spawned data-processing techniques and programs enabled airlines to provide instant flight information and reservation systems. Insurance companies have used these techniques to improve their accounting and investment services. Thousands of other firms today routinely process transactions involving millions of items a day.

Faster finances: new retail-store and bank-credit systems have caused great improvements in the speed and accuracy of business transactions, credit okays, and inventory control. Such systems have evolved from the automatic checkout equipment for the Apollo moon flights which used some of the most complex and advanced computer systems in the world.

Perhaps one of the best examples is the transferral of Apollo guidance computer software and data-communications methods to computerized retail sales systems for department stores. Many of the largest chain store businesses in the country use them. There are more than 50,000 point-of-sale terminals connected to these systems. A typical system may have 1,000 credit authorization terminals in up to 200 stores.

What are the advantages over nonautomated methods previously used? They include a nearly 100 percent reduction in purchases on bad-debt accounts; a 75 percent reduction in fraud purchases; a 20 percent cost savings in payroll because fewer credit authorization employees are needed; and a 33 percent reduction in telephone calls. Computerized systems with point-of-sale cash-register terminals also provide improved inventory control; more accurate and faster sales transactions; more detailed merchandising information; and better sales data for management analysis.

NASTRAN, NASA's general-purpose digital computer program, is used by scores of industrial firms, universities, laboratories, and government agencies to help solve tough engineering problems. The system has been adapted by businesses for hundreds of applications, ranging from suspension units and steering linkages on automobiles to the design of new power plants and skyscrapers.

Apollo computer software now is used in electric power dispatch terminals at many large public utilities. This provides for more efficient electric power distribution during peak demand periods, and it lessens the likelihood of blackouts or brownouts in highly populated regions.

Obviously, the march of computer technology speeded by space-program needs is now enabling business and industry management to make sounder decisions, and often better products, on the basis of faster, more accurate, and more complete information.

The same is true in the fast-growing world of microelectronics. Developments here have been pushed ahead by years because of the stringent weight and volume requirements of spacecraft. Thousands of circuits now are compressed into minuscule chips not much larger—and a lot thinner—than a matchhead. These miniaturized components have hundreds of uses in commercial products. Electronic calculators no bigger than a pack of cigarettes are just one example.

Other spin-off developments are being used in business and industry for a wide variety of purposes. An improved inorganic paint developed by NASA as an anticorrosion coating has widespread potential application in industry. Part of the Golden Gate Bridge in San Francisco has been painted with it. Unlike other inorganic paint, which is difficult to apply, this potassium silicate zinc-rich coating sprays easily, features long protection, and doesn't require a finish coat. It provides corrosion protection from salt spray, fog, heat, and the thermal shock of rapid temperature changes.

Here are some other representative samples of space-to-industry spin-offs:

Vibration sensors that measure the effect of traffic and weather on bridges, determining structural integrity.

Electronic strain gauges that have a unique capability—not previously possible—to accurately predict meat tenderness after cooking. They are used widely by national meat packers.

Precipitation-hardened steel alloy, used in new wind machines for frost protection in apple, almond, and other noncitrus orchards.

Composite materials data, used to help develop new, improved products such as golf-club shafts and business machines.

To demonstrate the application of aerospace knowledge to advance the building industry in residential home construction, NASA has built a "Technology House" at its Langley Research Center in Virginia. Materials, tools, systems, management, and construction techniques used to build the house all are spin-offs from the space program. These include solar and wind energy collectors and converters, water and waste treatment systems, and advanced systems for heating, cooling, water heating, and lighting, among others—all incorporating the latest technology.

Satellites are being used in advancing the science of crime prevention and detection. The Federal Bureau of Investigation (FBI) is one agency that has availed itself of satellite services. The FBI operates a nationwide computer/communications system called the National Crime Information Center (NCIC). It serves local, state, and federal law-enforcement groups in fifty states and Canada through more than one hundred control terminals. NCIC contains more than two and a half million records on wanted persons, identifiable stolen property, and stolen vehicles. The system handles a daily average of well over 50,000 transactions.

The Bureau also has a computerized criminal history file for use in the NCIC system. Demands for this type of information are greatly increasing. Research studies by NASA and the FBI have proven that satellites are the best means of communications for speeding such criminal data to the control terminals, no matter how remote their location.

Advanced electronics techniques developed at the Johnson Space Center are being adapted into a new means for the high-speed, preliminary classification and identification of bullets. Like fingerprints, bullets have distinctive markings. They are made when the bullet is fired by the lands and grooves in the gun barrel. Forensic scientists usually identify or match a bullet with a particular gun by using a comparison microscope. This traditional technique, which has been in use since the 1920s, is a slow, methodical process. The new system, however, stores individual bullet "signatures" in a general purpose computer, and correlates a new signature with those on file in the computer library. This promises a great leap forward in the science of ballistics.

All these crime-deterring products, techniques, and systems are either in everyday practical use now, or are in the final stages of testing. In the future, even more striking law-enforcement spin-offs are forecast. For instance, scientists are talking of deploying a huge complex of mirrors, nearly a square mile in diameter, in earth orbit. Each of these complexes would illuminate up to 36,000 square miles of land surface through solar reflection. And the area illuminated would be about ten times the full moon's brightness on a clear night. Yet this would involve no power consumption and it would cause no environmental pollution. Such illumination systems undoubtedly would have a far-reaching effect on cutting down city crimes on once-darkened street corners.

6 TRANSPORTATION TRANSPLANTS

Rockets and spacecraft today travel at unprecedented speeds—thousands of miles an hour—as they whirl around the earth or zoom to distant planets. Some of the advanced technology developed to make such super-swift flights possible is readily adaptable to the improvement of transportation systems on earth. The use of aerospace computerized techniques, and microelectronic systems, among other things, is, in fact, revolutionizing sizable segments of the transportation industry—on the highways and railways, at sea, and in the air.

One of the most visible applications has been San Francisco's much publicized Bay Area Rapid Transit (BART), which has been called the nation's most modern municipal rail transportation program. BART links more than thirty stations in three counties by means of a 75-mile system that includes the world's longest underwater transit tunnel—beneath San Francisco Bay. BART's builders borrowed space electronic and computer technology, along with a systems engineering approach.

U.S. railroads will install, by 1980, a national freight car information system called TRAIN II. This computerized project will improve rail car use and help forecast future car needs by updating information such as car location and status on an hourly basis, as compared to the present daily basis.

A major aerospace company has designed a computerized dispatching system for railroads—one of the most sophisticated train control systems in the world. First employed by the Southern Pacific Company, the system uses television displays to provide a central dispatcher with the continually updated status of rail switches and trains.

Similar computer techniques have been used to develop the first computerized traffic control system in the United States. Called SAFER (Systematic Aid to Flow on Existing Roadways), it was initially put to use in a heavily populated area just south of Los Angeles. There, it produced a reduction of more than 15 percent in the millions of vehicle hours spent waiting at 112 traffic lights. SAFER systems also have been installed in Baltimore, Maryland, and Overland Park, Kansas, among other areas.

In still another joint program, NASA and the Federal

Highway Administration are developing a system called Randomec, which is used to detect structural deterioration in highway and railroad bridges. Currently, there is no reliable, inexpensive way to detect bridge damage that may have been caused by corrosion, metal fatigue, collisions, fire, or extremely heavy wind gusts. The most common detection technique has been visual inspection—but defects, strains, or flaws may be hidden by paint or cover plates. Randomec is based on a mathematical analysis method used to detect failure due to structural fatigue in aircraft and rocket components.

NASA's computerized Structural Analysis Program (NASTRAN) is being broadly used in the automotive industry. The Ford Motor Company, for one, applies it for design analysis of car, truck, and farm tractor components. Design engineers at General Motors also use it. Chrysler, too, is using aerospace electronics and computer technology to develop new products, and for production line testing.

Perhaps one of the most intriguing space-to-automotive benefits is a new studless winter tire with low-temperature pliability. It provides traction equal to or better than studded tires on slick snow or icy surfaces. It came about as a direct result of technology used to develop the tires for the mobile equipment transporter astronauts drove on the moon.

While these and many other innovations are in effect today, there is even greater promise of future technology transfusions for transportation ends. Consider the possibilities of a full-time commercial navigation satellite

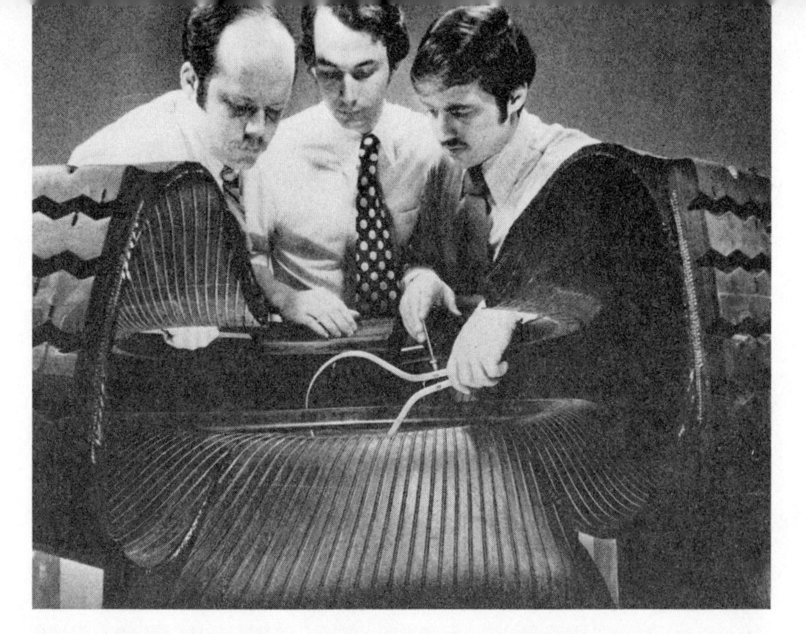

Studless winter snow tires have been developed by American industry using techniques first designed to help astronauts pull a mobile equipment transporter on the moon. Many states have banned studded tires because they can be harmful to roads. These studless tires, however, provide ample traction on icy roads without destroying the road surfaces.

system. It could lead to the day of the automated ship. Such a ship would leave port with a greatly scaled-down crew—mostly electronics specialists who would operate and maintain computers and other specialized equipment. Position information relayed by satellite would be fed directly into the ship's computers, enabling it to stay on a perfect prescribed course to any destination in the world.

Satellite signals could be received automatically, giving captains their sea positions rapidly and continuously. And the system would operate under all weather conditions, even during bad storms. A pipe dream still decades away? Hardly.

At present, navigation satellite systems are too expensive for commercial use. But the day is soon coming when these costs will be brought within reach of shipping companies. It has been estimated that the maritime shipping industry could save hundreds of millions of dollars a year through "optimum-time-routing" of ships, improved scheduling of ship arrivals and unloading, and more efficient use of personnel and equipment, if they could improve their ability to route, schedule, and predict ship arrival times.

Satellites also offer great potential in air-traffic control. Aerosats, as they may be called, could provide operational communications, such as airline message traffic, navigation, traffic control, collision avoidance, passenger telephone service, weather advisories, and search and rescue data.

Space experts believe these types of satellites will one day be able to establish the location of aircraft or ships within a positional accuracy of sixty feet or less. They will be able to operate continuously, regardless of weather conditions.

Thus, when such new satellite systems are employed, perhaps in the 1980s, they will substantially reduce the aircraft and ship accident rate, and increase the efficiency of transoceanic and coastal transportation of both freight cargoes and people.

But the ultimate transportation spin-off may well be rocket-powered air travel. Technology already available will one day produce a space transport system that could carry passengers from New York to Tokyo in forty-five minutes, or from Los Angeles to Rome in forty minutes.

When NASA was created by an Act of Congress in 1958, it absorbed a smaller organization that had been functioning since 1915—just twelve years after the Wright brothers' first flight at Kitty Hawk. This was the National Advisory Committee for Aeronautics (NACA). Today, the Space Agency is continuing efforts to meet some of the basic objectives charted for NACA more than sixty years ago. These efforts, over the decades, have contributed a major share in the building of the United States' position as world leader in the development of civil and military aircraft.

NASA's present and planned aeronautical research programs are targeted at developing quieter, safer, and more comfortable aircraft that are more economical, conserve more fuel, reduce the impact on the earth's ecological balances, and provide better public service. Toward the accomplishment of these broad goals, NASA scientists, engineers, and test pilots work closely with other government agencies and private aircraft industries.

Some of the most important work involves fuel conservation. This is especially true today, in an era of soaring fuel costs, diminishing supplies, and the ever-present threat of oil embargoes. Through a concentration of efforts, NASA has identified and is developing specific technologies—both near-term and those that can be attainable in ten years—which have the potential of reducing commercial aircraft fuel requirements by as much as 50 percent. This could mean savings of hundreds of millions of barrels of oil per year.

How? Through reduced air resistance, or drag, reduced weight, more efficient engines and control systems, new design concepts, and better operating procedures—both in the air and on the ground.

For example, studies have proven that a "supercritical wing" designed and built by NASA, effectively reduces both aircraft drag and weight. Featuring an entirely new design, this wing has been successfully tested on jet aircraft, and increases fuel efficiency by about 15 percent. This innovation alone will enable planes to fly faster and farther without increasing engine thrust or fuel consumption. It also will allow aircraft to carry more cargo over a set distance with the same amount of fuel.

Computers are playing a big part in advancing the aeronautics art, too. A new flight-control system, called digital fly-by-wire, is both fast-acting and computer-coordinated. Its use on aircraft improves handling qualities and improves flight safety. It also helps reduce aerodynamic loads and structural weight. With it, lightweight wires replace the heavy system of metal rods, hinges, and hydraulic lines that previously translated pilots' signals from cockpits to aircraft control surfaces. NASA estimates that aircraft modified with this system could fly about 15 percent farther on the same amount of fuel.

Work is also being done on composite material applications to aircraft. These strong, lightweight combinations of metals and plastics, such as boron or graphite epoxy, can reduce airplane weights by 30 percent or more when compared to aluminum structures. This could save another 10 to 15 percent in fuel consumption.

With protecting the environment a primary concern in new aircraft design and production, much NASA research on jet engines is directed toward helping to decrease substantially the noise levels of jet planes.

A lot of aeronautical effort is being put into development of V/STOL (Vertical and Short Takeoff and Landing) aircraft, which offer great promise as safe, clean, quiet, and efficient jets. They would require much shorter runways, because they would combine the takeoff and landing features of both a helicopter and an airplane.

Commercial development of V/STOL craft could virtually eliminate the heavy traffic congestion that today plagues major airports. With their introduction, present large metropolitan jetports could be used almost exclusively for transcontinental and transoceanic airliners. Smaller, less crowded V/STOL airports could be used for relatively short intercity flights of less than five hundred miles. These flights now account for about 80 percent of all air travel, and they are expected to triple in the next few years.

On many of these flights it takes nearly as long or longer to get from the airport into or out of a city, a few miles away, as it does to fly between cities several hundred miles apart. Smaller, specially designed V/STOL "Quietports" built nearer to city centers would solve this problem.

There are scores of other aeronautical studies underway; to name a few, drag brakes for steeper approaches, more efficient air and runway traffic control, and high-capacity landing gears for quick runway turnoffs. Several such programs are covered in this book in the chapter on safety spin-offs.

Through such activities, the application of space technology is continuing to improve air transport. The benefactors include travelers, shippers, operators, and all others whose lives are affected by the airplane.

7 NEW IDEAS IN ENERGY

The accelerated onrush of technology that has mechanized, automated, and computerized our world, mostly within the past century, has made life easier, safer, better than ever before. But along with the mainstream of progress all this technology has fostered, there have been some disadvantageous side effects.

One is the waste by-products of a highly industrialized society, which take various forms of environmental pollution. Another is the enormous amounts of "food" required to keep such a society in motion. Especially in the past few decades, there has been a constant and enormous drain on the earth's limited reserves of such essential fuels as coal, oil, and gas used to create energy.

It took millions of years for natural processes to produce these reserves; yet we have been burning them up at such alarming rates that there is concern that the still-existing stores will be exhausted within perhaps a few decades. Only in recent years, really only in the 1970s, have we begun to recognize this fully.

The United States and other countries throughout the world are now directing efforts to discover new fuel reserves and to develop new energy sources. Again, spinoffs from the U.S. space program offer some of the most promising answers for solving the problem, both for the near-term and for the longer-term future. NASA, in fact, has been working on energy improvement programs for more than fifteen years. Initially, research was directed for space flight uses, but now much of this accumulated knowledge is being channeled toward commercial outlets on earth.

The Landsat satellites, for example, already are at work searching from their orbital vista points for hidden and undiscovered geological formations that are providing clues to possible new sources of petroleum.

Landsat, or future-generation satellites like it, also may be used to find geothermal sources of power, which lie beneath the planet's surface. These sources form the heat generation that causes such natural phenomena as geysers and volcanic flows. Scientists believe if these geothermal pockets can be pinpointed, studied, and better understood, they may eventually be harnessed for everyday power uses. NASA's Jet Propulsion Laboratory in Pasadena, California, is now working closely with the Energy Research and Development Administration to set up a master plan for the effective development and use of geothermal resources on a national and regional basis.

Ironically, in the broad-ranging search for alternate means of creating electrical power, one of the most ancient systems is being reinvestigated—the use of windmills. Wind energy is a clean, limitless energy source that has, for centuries, been used for everything from pumping water to grinding grain. Because windmills could not compete economically with fossil fuels, few such systems have been used in recent years. But with oil and gas supplies rapidly dwindling, and their costs sharply rising, wind energy may again become competitive.

It is estimated that winds in the Great Plains area of the United States alone could supply about half of the growing U.S. electrical requirements for years to come. Windmills can directly spin generators for production of electricity. Or air can be compressed and later used to drive air turbines that would operate generators. The storage system provides available energy for use when the wind dies.

Consequently, a joint NASA-Energy Research and Development Administration program is aimed at applying space spin-off technology in aerodynamics, structures, materials, and power generation—to new windmill design. Much of the research work is being done at NASA's Lewis Research Center near Cleveland. And practical applications are already being envisioned.

The ultimate energy source, of course, is the sun. Nearly all energy available to us is, or was, solar energy

One of the oldest and cleanest forms of providing energy is the windmill. NASA scientists are studying this vertical-axis windmill for the conversion of wind power to electricity. The airfoil- (wing-) shaped blades rotate in almost any wind to provide the energy requirements of a typical single-family house.

in some form. With an estimated five billion more years of life in its present state, the sun is a virtually inexhaustible source of fuel. And direct applications of its power are virtually pollution-free. NASA is now directing a concentrated effort on developing ways to tap the heat and light of the sun to meet mankind's future energy needs.

Probably the first large-scale direct use of solar energy will be in the heating and cooling of homes. About 25 percent of current fuel consumption in the United States

NASA is assisting federal energy agencies in a number of windmill study programs designed to provide a clean and inexpensive source of energy. This experimental windmill reaches maximum kilowatt output in winds less than 20 miles per hour, and can generate enough electricity to power about thirty homes.

is for heating and air-conditioning homes. Direct sun heating is based simply upon absorption of solar radiation. This is done through what is called the "greenhouse effect." That is, glass is transparent to most of the sun's rays that reach earth. But glass is opaque to reflected or emitted heat wavelengths from objects under the glass. So glass acts as a one-way valve, admitting light but retaining the resulting heat, or at least a part of it.

New solar collectors are being designed and installed that cut down on radiation losses due to convection as heat rises from the absorbing surface to glass panes; by conduction of heat through the panes; and by radiation from the hot absorbing surface.

The solar collecting systems essentially work this way: air, water, or some other fluid passes through the collector—behind or through the absorbing surface. In the winter, the fluid transports the collected thermal energy, or heat, to the building heating system. In the summer, it is transported to a heat-powered refrigerator to supply cold water or chilled air for air conditioning. The fluid also may convey thermal energy to an insulated tank where the heat is usually stored in water, rocks, scrap iron, or pieces of concrete until needed for nighttime use.

Today, a number of large buildings, and a few homes, in the United States are being heated partially with solar collector systems. Also, some homes in the southern United States, Japan, and the Middle East have solar water heaters. Water heated by the sun is stored in huge tanks. The lack of well-engineered, reasonably priced systems has been the key reason for such limited use of solar power to heat homes to date. However, the matchup of new technology—much of it developed for space programs—to increasing energy needs will soon change that.

NASA now is testing and gaining practical experience in direct solar heating and cooling by means of home systems engineering demonstrations. At the Langley Research Center, for example, a 15,000-square-foot experimental collector system has been added onto a one-story office building. Several different solar designs are being evaluated here. Through these test programs, NASA hopes to halve the cost of current direct solar heating and cooling systems, and to increase their lifetimes to beyond fifteen years. All of the knowledge being gained through these pioneering efforts is being made available to commercial builders and others interested in developing solar systems to heat and cool homes, offices, and other structures.

Another program that holds high promise for widespread future energy-producing uses on earth involves solar cell arrays. Conversion processes always lose or waste energy. Most processes for generating electricity from solar radiation convert it first to heat, which drives an engine that in turn spins a generator. But solar cells can convert light—even on cloudy days—into electricity without moving parts.

Eventually, scientists envision vast solar cell arrays that could be set up in sunny areas such as Southern

California, Arizona, or Florida, and be used as "energy farms." Solar power would, of course, be cut off during the night, but then stored energy could be used.

One of the most exciting ideas being studied as a future energy source is the satellite solar power station. A satellite in synchronous orbit above the earth's equator receives solar energy for almost twenty-four hours a day. In this orbit, a satellite receives six to ten times the amount of energy that can be collected by conventional systems of the same size on the ground. The power generated in space could be beamed via radio waves to the ground, where it would be converted to a form useable for everyday needs.

It will be a number of years before satellite solar power stations can be developed for duty in space, possibly the 1990s or later. Other systems will not take as long. For example, solar heating systems could be in relatively heavy commercial production within the next two to five years. Solar heating and cooling systems will take a little longer—five to eight years.

However long it takes, there is a consensus among scientists that solar-powered systems represent the greatest long-term potential source of energy for a world that is fast running out of fossil fuels.

Thus, the broad-ranging use of space technology is helping speed the day when people on earth will no longer have to rely on the limited and increasingly expensive resources of their planet to provide the energy that is essential for a future of continued progress and growth.

8 ONE-WORLD COMMUNICATIONS

"Live via satellite."

These three words, now taken for granted, have shrunk our planet by revolutionizing the world of communications. Such historic events as the Olympic Games, overseas visits of the U.S. president or other heads of state, and major news occurrences of the hour are beamed instantaneously to television sets all over the world, thanks to the miracle of communications satellites in orbit. It is estimated that more than one billion people—one out of every four on earth—can see an international event on television as it happens, regardless of its point of origin.

Even so, television is not the principal benefit of these versatile spacecraft. An even greater payoff is being made in direct economic gain to world commerce.

A major portion of all long-distance international communications, and more than one-half of all trans-oceanic communications, are now being sped by satellite. The spatial networks are also capable of providing a variety of voice, video, data, and graphic transmission services to nearly all fifty states and to more than one hundred nations overseas. Means-of-communications traffic routed through space includes data and facsimile transmission, teletype, and even Mailgrams. And all this has developed since the first transoceanic television transmission via space less than two decades ago—on July 10, 1962.

Artificial satellites were first launched into earth orbit in 1957. They are hurtled into space, anywhere from about one hundred miles to several thousand miles above the planet, by powerful rockets working on the principle of Newton's third law of gravitation. That is, to every action there is an equal and opposite reaction. Rockets rise off their launching pads because their fuels are blasted out through a rocket nozzle at high speed, just as a child's inflated balloon zips through the air when the rubber band is taken off its neck.

The rocket's motors cut off as it reaches orbit above the earth, and the satellite is popped out of the rocket's nose. The satellite then continues to move, obeying Newton's first law: a body in motion (in the absence of friction) continues to move at constant speed in a straight line unless it is acted upon by another force.

In the near vacuum of space there are not enough air molecules around to cause appreciable friction. Therefore, the satellite maintains a course circling the earth, high enough to be free of the pull of earth's gravity. It can stay on such a course for years, traveling around earth in much the same manner as the earth orbits around the sun.

The first artificial satellites were tiny, weighing only a few pounds. But as rockets and their fuels were improved, heavier loads could be launched into space. Today it is common to boost satellites weighing several hundred, or even thousand, pounds into orbit. Most satellites are cylindrical in shape, have protective metallic exteriors, and are packed with miniaturized electronic instruments.

Satellites offer unique advantages for communications purposes, unmatched by such conventional systems as submarine cables, land lines, and microwave radio stations. Because microwave does not bend and travels in a straight line, relay stations must be placed every thirty miles or so to allow for the earth's curvature; thus they can receive, amplify, and retransmit signals. Submarine cables have been used to span continents, but they, too, are very costly to install and maintain. They are often severed by ships, and generally have had limited capacity. Microwave and cable also are subject to the elements—wind, rain, and any other natural disturbance that can interrupt service without warning.

A single satellite, on the other hand, placed in a synchronous orbit above the earth at the equator, remains in a fixed position where it provides complete communications relay coverage for 40 percent of the earth's surface, across an ocean to two or more continents. A synchronous orbit is one in which, at an altitude of 22,300 miles, the satellite's orbit matches the rotational speed of the earth. The satellite appears to hover continuously in one spot. Three such satellites, spaced over the Atlantic, Pacific, and Indian oceans, at 120-degree intervals about the equator, can effectively cover virtually the entire globe. Satellites are constantly in these positions. When one goes "mute" electronically, another is sent up to replace it.

These satellites have communications systems consisting of receivers, amplifiers, and transmitters. Signals received from stations on earth on one frequency are amplified and transmitted on another frequency to other earth stations. This makes it possible for a ground station to transmit and receive at the same time. For example, transmission of a televised event such as a football game in the United States can be beamed from the ground to a satellite positioned over the Pacific Ocean. The satellite then can relay it for reception in Japan or other Far Eastern countries. At the same time, an event occurring in Japan can be seen in the United States by operating through the same spacecraft.

Earth stations, then, serve as the connecting links between land communications systems in the countries where they are located, and the satellites. Large antennas are used to relay all forms of communications. This is how land and space facilities are combined to relay telephone, television, telegraph, data, and facsimile communications between distant points at the speed of light.

Also, since satellites have a multipoint communications capability, stations in many countries can communicate with one another at the same time.

The U.S. Communications Satellite Act of 1962 provided for the creation of a new, privately owned corporation to be the U.S. agent for participation in the system. This was the Communications Satellite Corporation, or Comsat. In 1964 the International Telecommunications Satellite Consortium—Intelsat—was formed, with fourteen nations as charter members. Today Intelsat has more than ninety member nations—from Afghanistan to Zambia. Each nation jointly shares the costs and services of the satellite operation. The Intelsat network of earth stations has more than 115 antennas at over 90 station sites in more than 65 countries.

Antennas protrude like giant electronic dishes from the dense jungles of South America, the arid deserts of North Africa, the deep snowbanks of Northern Alaska, and even from the tops of giant oil-pumping rigs in the North Sea. And hundreds of additional antennas are springing up around the world. Indonesia is building a network of 60 to link its more than 2,000 islands by satellite. Micronesia, Algeria, the Philippines, and Brazil are working on extensive systems of antenna networks.

The first commercial communications satellite for Intelsat was Early Bird I, launched in April 1965, and placed in orbit over the Atlantic. It had a capacity for handling 240 telephone conversations or one television program.

In contrast, today's advanced, drum-shaped Intelsat

IV-A communications satellites each carry 7,000 telephone circuits. They can transmit up to 20 simultaneous color television programs from continent to continent, spanning the world's major oceans. Or they can route any combination of "traffic," including data and facsimile, or tens of thousands of teletype circuits.

Each Intelsat IV-A has an orbital life expectancy of seven years. Such relative longevity and the greatly increased capabilities of this new generation of satellites has meant not only better and more easily available service for users, but also less expensive rates.

In November 1972 the world's first domestic communications satellite was launched—Canada's Anik 1. A second Anik (the Eskimo word for "brother") was orbited six months later. Together the satellites provide, for the first time in Canadian history, the means for total transcontinental communications, instantly and efficiently. Previously, the only way of making contact with hundreds of widely scattered, sparsely populated villages in remote, frozen northern tips of provinces was through shortwave radio, which is unreliable during frequent periods of severe weather.

The first American domestic communications satellite—Westar 1, launched for the Western Union Corporation—was orbited April 13, 1974. A second Westar was placed in orbit several months later, as have other spacecraft, such as Comstar for the American Telephone and Telegraph Company. These craft can provide a variety of voice, video, data, and graphic transmission services throughout the fifty states. Within

Westar, the United States' first domestic communications satellite, was launched into orbit by NASA from the Kennedy Space Center in Florida in April 1974.

the first few months of Westar's launch, more than four hundred government and commercial users, including large and small companies, had ordered satellite voice/ data circuits.

In September 1974 the first mail was sent by space. Thousands of special Mailgrams were transmitted from an earth station in Glenwood, New Jersey, to Westar 1 in orbit and back down to Steele Valley, California—a total distance of 47,000 miles, yet they traveled as no other mail ever had: at the speed of light.

Late in 1975 a major newspaper, *The Wall Street Journal*, started daily production through the use of high-speed facsimile transmission via satellite. Full pages of advertising and editorial copy are electronically transmitted from the paper's printing plant at Chicopee Falls, Massachusetts, through space, to a new plant in Orlando, Florida, where it is received on page-size photographic film and then readied for printing. The total transmission time is little more than three minutes a page.

One of the most recent advances has been the design of a new maritime satellite system called Marisat. This provides communications service via satellites over the Atlantic and Pacific Oceans to ships of the U.S. Navy, commercial ships, and to offshore industry.

The spin-off benefits from space in the field of education have been both dramatic and plentiful, and, in fact, are helping wipe out illiteracy and ignorance all over the world. This has perhaps best been exemplified through a series of NASA experimental Application Technology Satellites (ATS). The first three, launched in the late

Engineers and technicians prepare a Marisat spacecraft for launch. Marisats provide communications service via satellites over the Atlantic and Pacific Oceans to ships of the U.S. Navy, commercial ships, and offshore industry.

1960s, quickly demonstrated vast potential for improving the quality of education through the use of space.

The first spacecraft in the series was used in an experimental program in Alaska. In one phase, about twenty-five remote communities in the state participated in a series of public radio broadcasts made possible by ATS 1. Teachers at remote, outlying districts conferred by two-way voice communications via satellite with professors at the University of Alaska to advance their teaching skills.

But by far the most exciting results to date are coming through an improved spacecraft in this series—ATS 6, launched in mid–1974. Instrumentation aboard it adds the new dimension of video to small terminals. Here are some examples of educational advances being made possible by ATS 6:

It provided videotape courses and live seminars to more than 1,200 teachers in the Appalachian regions of Virginia, Alabama, Tennessee, Maryland, and New York. Programs included pre-taped television instruction and live seminar broadcasts in addition to group discussions. Many of these educators, in mountainous areas far from large cities, have had little access to such higher education projects. The use of space technology is changing that.

In Alaska, eighteen communities scattered across the state joined in two-way classroom instruction sessions. None of these communities had previously had television reception of any kind. Here, ATS 6 was also

Artist's concept shows a NASA Application Technology Satellite (ATS) as it appears in space. These versatile craft have demonstrated vast potential for improving the quality of education through the use of space by beaming educational programs to remote areas of the world.

used for telemedicine experiments. Doctors in remote outposts consulted directly with top specialists, transmitting patient's records and X rays, and actually examining the patients before television cameras in live, two-way voice and picture transmissions.

Ten Veterans Administration hospitals in the continental states used the satellite to provide live, interhospital exchanges of medical data, patient case studies, and in-service training seminars for doctors and medical personnel.

Following more than a year of such service over the United States, ATS 6 was repositioned in space at a point over Lake Victoria in East Africa. From here it is in "line of sight" with an experimental ground station in

Technicians of the Indian Space Research Organization check out a direct-receive antenna installed in the tiny village of Kerelli, about 300 miles southeast of Bombay. It is part of a network of antennas set up to receive educational programs in remote villages from NASA's Application Technology Satellite.

Ahmedabad, India. The Indian government transmits educational television programs from a few transmitting stations to ATS 6. The first programs are on two subjects of paramount interest to India—population control and agricultural improvement. The satellite then retransmits these programs to hundreds of thousands of people in five thousand rural villages, each equipped with small television receivers.

Eventually, more than half a million direct-broadcast receivers will be positioned in Indian villages, serving audiences of up to several hundred people at each location. The cost for this system is about half of what it would cost if it were ground-based.

Many educators agree that the space endeavor has steadily upgraded the educational system in America.

Consider, as just one example, how space flight has changed our thinking about the moon, Mars, and other planets in the solar system, including earth. The view astronauts had of earth from the moon gave us a whole new perspective, altering our concepts about our planet's environment and resources. Courses in everything from astronomy to geology had to be restructured and reformed to be brought up to date. The same was true with mathematics. The curricula of many school courses have been vastly improved through a substantial revision of content, a reorganization of subject matter, and fresh approaches to the studies. And one day in the not-too-distant future, satellites will broadcast radio and television programs to people in all countries on all continents, a truly global educational system that will improve school courses around the world. Such a system will eventually permit thousands of programs to be broadcast simultaneously on different frequencies. Viewers will be able to select and receive the best and the latest information on whatever subjects they choose. By 1985, the quality of education should no longer depend on the locale where students happen to live.

Experts believe that the day will come, possibly in this century, when universal satellite educational programs will lead to the learning, and the understanding, of a common language by all the world's peoples.

9 SATELLITE WEATHER WATCH

In August 1969 a monstrous killer hurricane snaked its way across the Gulf of Mexico. It lashed into the Louisiana–Mississippi Gulf coastline with a fierce strength and intensity unlike any other storm of modern times. Howling winds of more than 200 miles per hour drove great walls of water across the low-lying beach communities. The unrelenting water, twenty-four feet above mean sea level, literally flattened entire towns, smashing homes, stores, buildings, piers, and marinas. Hurricane Camille, and the widespread flooding that closely followed its devastating path, caused a record $1.5 billion in damage, and left tens of thousands of people homeless.

Remarkably, only about 320 people lost their lives, and more than 100 of these people drowned in the floods

A tropical storm as seen from space: Hurricane Katrina, packing swirling winds approaching 100 miles per hour, was photographed off the coast of Baja California by a Synchronous Meteorological Satellite. Tracking such storms from space has enabled early warnings to residents in their paths, saving thousands of lives and hundreds of millions of dollars in property damages.

after the hurricane had left. A storm of Camille's size and power even twenty-five years ago would have killed tens of thousands.

But thanks to a modern miracle made possible by the space program, warnings of the approaching hurricane were flashed early enough to enable 75,000 residents of the coastal region to evacuate to safer, higher ground.

Refining the science of weather forecasting, a government meteorologist prepares weather maps from satellite photographs.

The warnings came via a meteorological satellite, which, from its all-seeing vantage point in earth orbit, could track Camille, and relay pictures of the storm's progress to forecasters days ahead of the time it struck the coast.

Today, weather satellite photographs are routine, commonplace. They are seen daily on local television newscasts. But they are relatively new. NASA launched its first meteorolorical test satellite in 1960, and it has only been in the past decade that such spacecraft have been placed on an operational status.

The first weather satellites were called Tiros (Television and Infrared Observation Satellite). In all, ten Tiros spacecraft were orbited in five years, pioneering a new era in weather observations and forecasting. One of the most important contributions of the entire program was the development of a revolutionary new camera system—the Automatic Picture Transmission (APT). APT, when used with relatively inexpensive ground receiving equipment, led to the general use of space data from satellites in the preparation of local weather forecasts all across the United States and in other countries.

Now hundreds of weather stations in more than one hundred countries all over the world have their own APT receivers—and immediate access to the most up-to-the-minute weather information direct from satellites.

Following the first Tiros flights, NASA launched a larger, more sophisticated spacecraft called Nimbus—the Latin word for "rain cloud." Nimbus television cameras and other instruments are capable of producing pictures of weather patterns in *both* sunlight and nighttime portions of the satellite's orbit—a major advancement.

The technological breakthroughs made by Tiros and Nimbus led, in 1966, to the initiation of the first full-time operational weather satellite system.

By the 1970s, a new, "second generation" series of spacecraft took over the weather watch from space. These were Improved Tiros Operational Satellites

(ITOS), which offered, for the first time, twenty-four-hour coverage of the earth on a routine basis. Cloud-cover pictures at night were made possible by a new scanning infrared radiometer. Infrared has become one of the weatherman's most valuable tools. Infrared sensing instruments in satellites can pinpoint the temperatures of the land, sea, and cloud tops to an accuracy of within 3 degrees Fahrenheit.

On the ITOS spacecraft, an attitude control system keeps the specialized instruments continually pointed at earth. Each satellite photographs a strip of clouds about 1,700 miles wide and 25,000 miles long every two hours. And each picture taken by the television camera systems covers an approximate 2,000 square miles, or about four million square miles of cloud cover. The satellites are programmed to pass over roughly the same points on earth at the same time each day.

The one technical shortcoming of the ITOS system is that while these satellites cover the entire globe daily, it takes a full twelve hours for the satellites to get back to the same spot again. They do not provide continuous observations of all points on earth. Thus severe weather, such as thunderstorms or tornadoes, could develop in an area after an ITOS pass, and cause widespread damage with little or no warning before being detected from space.

To help fill in this critical gap, a new series of spacecraft has been designed and tested and is now on duty in earth orbit. This is called the Synchronous Meteorological Satellite (SMS) program. It has been developed to

Readied for launch, this Synchronous Meteorological Satellite is prepared at the Kennedy Space Center in Florida. This satellite program has been developed to keep significant portions of the earth's cloud cover under constant surveillance.

keep significant portions of the earth's cloud cover under constant surveillance.

The Synchronous Meteorological Satellites have an environmental data collection system that receives information from thousands of sensing platforms placed at remote sites on land, in ships and in buoys at sea, in rivers and lakes. From space, the satellites interrogate these sites, gathering such information as amounts of

Spectacular view: southern portion of North America and most of South America can be clearly seen in this photograph taken by a Synchronous Meteorological Satellite 22,300 miles above earth. Tropical storms can be seen across the top of the picture from western Canada to the Atlantic Ocean. Fog appears just off the coast of California. Snow-capped mountains in South America stand out prominently.

rainfall, river and stream heights, wind conditions, air temperature, sea state, and even earthquake measurements and volcanic disturbances. Pertinent data then is relayed from space to small regional warning and forecast stations all over the world.

Yet another series of weather satellites currently is under development. Known as Stormsat, its principal job will be to report on short-term, small-scale weather phenomena such as thunderstorms, squall lines, tornadoes, hurricanes, flash floods, and winter blizzards.

Through the recent major advancements made in meteorological satellites and sensing equipment, experts believe it will become possible to predict reliably the world's weather up to two weeks in advance within the next few years.

What will this mean? In the United States alone, an accurate five-day forecast of weather conditions over the United States would provide an estimated annual savings of several hundred million dollars or more—when applied to agriculture, lumber businesses, surface transportation, retail marketing, and water and land resources management.

When long-term forecasts become available and are disseminated worldwide on a timely basis, the savings could amount to hundreds of billions of dollars a year. Advance knowledge on long-term trends and seasonal features, such as an abnormally wet or dry summer, will enable farmers to plan better for the growing season. Forecasts of an abnormally cold or warm winter would permit proper planning by the power, construction, winter sports, and other affected industries.

Even more important is the saving of human life that accurate long-range weather forecasts will make possible. Predicting the occurrence and location of floods, tornadoes, hurricanes, typhoons, possibly earthquakes, and other forms of severe weather will greatly reduce the tens of thousands of deaths that annually result from such phenomena.

10 ORBITAL PROSPECTORS

When historians of the future record the significant dates of the twentieth century, July 23, 1972, will be one of them. Few people today, however, recognize it as the date a rocket launched a new satellite called Landsat 1. Many scientists say that some day its development may compare with the invention of the wheel or the discovery of fire, so revolutionary and so basic may become its ultimate impact on human society.

Landsat is an unmanned spacecraft with instruments aboard to survey the earth from space as man has never before been able to do. It is a versatile, multipurpose satellite that has performed far beyond expectations in surveying and monitoring applications involving agriculture, forestry, and range resources; water and marine resources; land use and mapping; environmental

America's first Landsat satellite is prepared for flight. It was launched July 23, 1972, and began surveying the earth from space as man had never before been able to do.

surveys; mineral resources; and geological surveys.

A second Landsat was launched January 22, 1975. In their relatively short lifetimes, the two satellites have performed daily wonders while sweeping the earth's surface steadily with their unblinking electronic eyes. They have spotted probable sources of oil and mineral deposits, previously unknown, in several parts of the world. They have tracked forest fires and sighted potential earthquake sites. They have discovered lakes and other bodies of water that man did not know existed, and found that others actually were located miles from where they had been placed on maps. They can detect crop diseases from space before farmers can in the field. They can pinpoint large schools of fish in the sea, and isolate productive grazing lands surrounded by vast desert wasteland. They can follow, with unprecedented accuracy, the migratory paths of wildlife. They can survey, every eighteen days, the damage caused by floods, hurricanes, or other natural disasters. They can measure ice floes and snow depths in remote regions that man cannot reach.

And they can do these and countless other invaluable services faster, better, more accurately, more repetitively, and for much lower cost than any other system ever designed. They can watch, survey, and report on a vast range of phenomena that human eyes cannot see.

What does this mean to people on earth? These ubiquitous satellites will help man grow more and better food; find and develop essential minerals and energy sources that might otherwise never be found; and help him avert or reduce the losses caused by the forces of nature. Landsat provides the tools that are enabling man to manage intelligently the finite resources of planet earth.

The Landsats, butterfly-shaped craft each weighing a little more than a ton, operate from orbits 560 miles up. Each Landsat makes a complete orbit of the earth once every 103 minutes, or about fourteen such orbits a day. The satellites are oriented so they constantly aim their instruments at earth. From their orbits, they are able to see any point on the planet, except for small areas around the North and South poles.

The flight paths give each craft the capability of viewing the same spot anywhere in the world at the same time of day—once every eighteen days. Thus, with two

Striking: this first coast-to-coast Landsat photomosaic of the United States was made from 569 virtually cloud-free images taken by the satellite orbiting at a height of 570 miles.

Landsats, every specific site on earth is covered once every nine days. This allows scientists the opportunity to observe changes on the earth's surface almost as they occur. Each Landsat continuously views a 115-mile swath of earth, running north/south, as it orbits. On the average, about four million square miles of earth are surveyed daily by each craft.

The satellites survey through a technique known as remote sensing, which means acquiring knowledge about an object from a distance. Remote sensing is possible because everything—living or nonliving—emits, absorbs, and reflects, in its own distinctive way, electromagnetic radiation that can be detected by sensitive instruments. An object's spectral characteristics—that is, the distribution of reflected electromagnetic radiation—often is called its signature. These are as distinctive as fingerprints. Signatures not only differentiate objects, but they also can indicate sizes, shapes, densities, surface textures, moisture contents, and other physical and chemical properties.

For example, wheat has a different signature from that of oats or corn. Identification of these signatures can tell scientists not only what the object is, but how old and how healthy it is. The cell of a sick plant reflects or emits radiation differently from the way a healthy cell does.

So prolific is this satellite system that during the first six months of its operation, Landsat 1 "imaged" nearly 40,000 scenes of the earth's surface. From these, about one and a half million high-quality photographic images were made and sent from NASA to government agencies and investigators all over the world for use.

Landsat I "photo" of the Monterey, California, area provides an example for making accurate maps of geologic structures mainly on the basis of the numerous linear features displayed in this image. Such linear features often relate to geological faults, which reveal the dynamic processes taking place in the earth's crust, and are indicative of earthquake activity.

One of the most exciting facets of the Landsat program has been the satellites' capabilities, through processing of the images they take, to locate potential new sources of scarce minerals, oil, and other valuable resources, often in remote regions of the earth. By using Landsat imagery of areas with known petroleum deposits and correlating these with Landsat imagery of geologically similar structures elsewhere, there is a good possibility that geologists will be able to determine where new petroleum deposits may be located, say NASA officials.

Landsat is particularly useful in this kind of search because it has exceptional ability to show large structural features, such as major fault systems, domes and uplifts, and folded mountain belts of regional or subcontinental size. And petroleum exploration and oil companies are using such data to help them narrow down the hunt in all corners of the world.

One of the greatest benefits of prospecting by satellite is that from space potentially productive mineral source areas can be spotted in desolate, barren, or hostile country that would not otherwise be searched, such as the backside of remote mountain ranges, arctic areas, or in the midst of a vast desert. For instance, Landsat data suggests further exploration in the rich Northern Alaskan oil fields. Images of the Umiat oil field region, which contains some fifty million barrels of oil, hint that due to magnetic "anomalies," gravity contours, and fold axes of the earth's crust, areas to the north and east of Umiat may hold more oil.

Crustal linear features discovered in the Tanacross

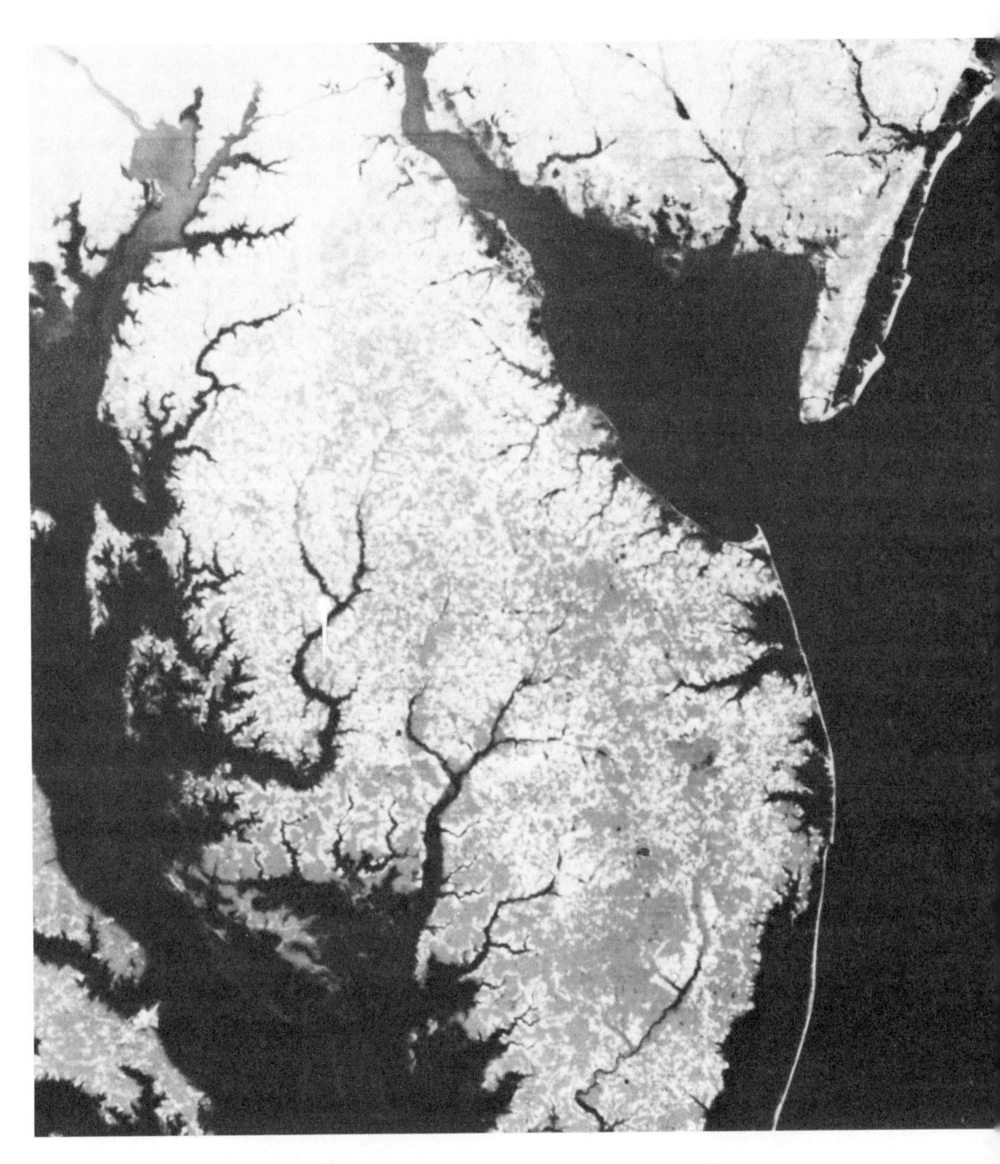

Upper Delaware, Maryland, and the Virginia Peninsula are graphically viewed in this composite Landsat photo. Chesapeake Bay is on the left; Delaware Bay at top center; and the Atlantic Ocean on the right. Such images are used by experts for land management, hydrology, meteorology, pollution control, oceanography, and other purposes.

area near the Canadian border, through Landsat images, have led to exploration for new copper ore deposits previously unknown. Prominent linears have been observed, presumed by experts to be major crustal fractures running northeast and northwest. Many known ore deposits are located close to such linears.

Landsats have found what appears to be the color and contour of nickel deposits in western Canada and in South Africa. Mining engineers are excited about satellite imagery that apparently shows two new large copper ranges in the isolated outer reaches of Pakistan. Scientists are following up on these and other tips relayed from space.

Another major application of Landsat data is in agriculture. Currently, Landsat 2 is involved in a large-area crop inventory experiment. The project combines Landsat imagery with meteorological data from weather satellites and ground stations to establish and refine procedures for predicting major crop yields. By singling out areas of food shortages and surpluses far enough ahead, inequities in supply and demand might be smoothed out, soaring food costs curtailed, and hunger alleviated as a global problem.

From Landsat in earth orbit, crop species in fields of twenty acres or more can be identified with high accuracy. Various species such as winter wheat, spring wheat, barley, fallow and harvested land can be distinguished. The satellites can monitor crop growth during the year, helping farmers predict crop yields. Crop diseases can be detected early by satellite through the use of infrared, which picks up "readings" that can distinguish between healthy and sick crops.

Investigators using Landsat images of California's Imperial Valley inventoried more than twenty-five separate crops in 8,865 fields in just forty man-hours. The total area covered was close to half a million acres. Among the crops readily identified were corn, popcorn, soybeans, sorghum, oats, four different kinds of grasses, lettuce, mustard, tomatoes, carrots, onions, and alfalfa. The scientists distinguished between wet-planted fields, plowed lands, harvested fields, and bare soil in areas as small as ten acres. To have done this job without use of satellite would have taken hundreds of man-hours and a lot of hard work.

In addition to such surveys and inventories, Landsat images are being used to discover unused rangeland suitable for cultivation in various parts of the world. The images are revealing such attributes as good soil, adequate water supply, and freedom from possible erosion. Landsat, for example, found an extensive rangeland, previously unsuspected, on Alaska's Kenai Peninsula. The area might never have been found otherwise because it was in an unpopulated region 80 miles from the nearest road. The data is being used to determine whether to develop the area for potential grazing of cattle or to let it remain in its wilderness environment.

In addition to helping spot crop diseases from space, Landsat is helping control some of the pests that cause such diseases. In one case, California farmers used satellite information in their war on the cotton bollworm. The most effective way to check this pest is by timely uprooting of cotton plants. This calls for tight scheduling, the harvesting of cotton as soon as it has matured, and then promptly plowing under the plants. To do so, farmers must closely monitor status of crop growth. At a test site in California's Imperial Valley, Landsat investigators provided a completed inventory of crop status in about 16 man-hours. Traditional methods require about 128 man-hours.

On the other side of the world, in Saudi Arabia, similar techniques using photography from space have been employed to track the flights and likely destinations of devastating hordes of locusts that strip productive land bare.

Landsat is also proving to be a useful tool in the monitoring of breeding habits of migratory waterfowl and of some animals. Annual production estimates are used to predict fall flights of ducks, for example. Information then is used for establishing waterfowl hunting regulations. Satellite remote-sensing techniques hold considerable promise for the accurate and rapid assessment of waterfowl breeding habitat, especially in pond numbers and distribution.

Already, Landsat imagery is being used in Canada and Alaska to map the previously unknown distribution of the habitat favored by walruses. In fact, the imagery differentiates between this information and data on locations of ringed, bearded, ribbon, and harbor seals, all of which are important north country food sources for Eskimos and others.

To foresters, a satellite system offers many advantages. It is the only thorough practical means to mount a continuous watch over large wooded areas . . . to provide warning of insect infestations and diseased trees . . . to take census of tree types . . . and to report logging yields. Instruments from space can sight and pinpoint, instantaneously, forest fires before man discovers them on the ground.

In another successful experiment, Landsat imagery was used in the Canadian province of Saskatchewan to map forty-two forest-fire burns across the remote northern part of the province. In the past such work was done by helicopters at a cost of $250 an hour, and would have run to a total of $10,000. The cost by satellite was less than $100, involved less manual time, and greatly improved accuracy in the mapping.

In Alaska, Landsat imagery has been used to develop a map of the major ecosystems of the huge state. Identified in the pictures are specific regions of very tall evergreens, hemlocks, Sitka spruce forests, high and low brush areas; forest tundra and coastal marsh; barren and sparse dry tundra; glaciers and ice fields; and freshwater areas from lakes and rivers. Similar benefits are being reaped in other parts of the country. In Georgia, Landsat imagery of the coast is providing spectacular definition of the Sea Island section and intervening marshland. These marshes are recognized by biologists as among the most productive nutrient systems on earth. Rich suspended nutrients washed from the marsh are carried to the fishing and shrimping grounds of the open ocean by a system of currents not previously well known. By studying the imagery from space, scientists are unlocking long-held secrets of nature that will lead to improved conservation measures.

The great Okefenokee Swamp, also in the south, is the largest elevated warm swamp on the North American continent, and one of the most prolific of the nation's wildlife preserves. But the origin of the swamp has never been well understood. Photographs developed from space imagery are giving scientists a new perspective to their studies of the swamp, providing a new tool to help measure the effects of man's continuing encroachment on the streams, marshes, swamps, wildlife, and plants of the area.

More than 70 percent of earth is covered by some form of water. Hydrologists are finding scores of practical applications from space information to the important field of water-resources monitoring and management. Such applications include snow-cover mapping, surface-water mapping, watershed surveys, estuary and wetland surveys, flood-area assessment, flood-plain mapping, river monitoring, and water-quality surveys.

In Texas, satellite data has been used in the mapping of thousands of lakes, reservoirs, and other water bodies, and in the inventorying of dams. Results show that all lakes larger than 40,000 square meters (10 acres) can be identified from space with 100 percent reliability, and their locations pinpointed to within 1,000 feet.

In Brazil, studies of satellite imagery found some large lakes in remote jungle areas as much as twenty miles away from where they had been previously mapped. In the Fiji Islands, a hydrographer used Landsat data to distinguish subsurface reefs and small atolls, and found that many coral atolls were out of relationship with neighboring atolls and islands. One atoll was found two miles out of its charted position. Such information is very important to shipping concerns, which run the danger of smashing vessels on hidden reefs.

Through Landsat, more rapid and complete surveys and assessments of drainage areas, stream network characteristics, vegetation cover and surface water features in large and remote areas are possible. This is being done in South America, Spain, the lower Mekong Basin of Vietnam, the Republic of Mali, and other locations around the world. In one specific case, covering a 170,000-square-mile region in South America, hundreds of drainage basins were inventoried, using thirty-one images from Landsat with reference to surface water features. Incredibly, thirty-six new lakes were discovered.

Landsat information also is being used in Pennsylvania to help locate areas of potentially high ground water yield; in Arizona to help investigators determine causes of grazing land erosion; in drought-prone areas of Florida to aid planners to better manage water resources; and in California and other states to help farmers better plan irrigation of their crops.

The versatile Landsats are proving useful, too, in flood monitoring. The U.S. Geological Survey studied satellite pictures of the Mississippi River Basin taken before, during, and after the devastating 1973 floods of that area.

By having a comprehensive picture of flooding over the whole river valley, scientists found that areas of maximum flooding extent could be determined, enabling the focusing of initial relief efforts on the hardest hit regions.

The U.S. Army Corps of Engineers today is using satellite imagery to inventory dams across the nation. Following the Buffalo Creek dam break in West Virginia in 1972, when many lives were lost and hundreds of homes were destroyed, the government ordered a large-scale inventory of dams and impoundments. Such a project would have taken years of work and millions of dollars to complete without the all-seeing eyes from earth orbit.

Many states are using satellite imagery to help in land-use planning and management. In Mississippi, the photographs from space distinguish entire transportation networks, including highways, interstate systems, primary and secondary road systems, railroads, power lines, and pipelines. Such information can be used in a wide variety of ways—from school-district planning of bus routes, to improved residential tax assessments.

Landsat also is helping to identify potential geologic hazards. It is leading to a new approach in deriving earthquake hazard maps. By plotting earthquake epicenters on Landsat images on which new linears have been identified, the degree of correlation between the earthquakes and surface fracture traces is being better substantiated. Already, data from space has been used to discover many geological surface features linked with earthquakes—particularly in sparsely populated western states—that experts had not known existed.

Under one NASA contract, for example, experts used blowups of specific land areas photographed from space, then charted on them all known faults and other geological features often associated with earthquakes. They carefully examined the photos for "hidden" land structures or other clues not previously sighted or known, and therefore not mapped. In studies of one large area covering a major portion of Nevada, experts found an astonishing 100 percent more faults than had been known to exist before.

Certainly one of the most unusual applications of Landsat imagery has been in the improving of mine safety. Coal mine cave-ins, for instance, are sometimes tied into local fractures or joints in the earth's surface. Some of these cannot be seen on the ground. Landsat has spotted several larger, through-running linears that are serious hazards, such as at the King's Station coal mine in southwestern Indiana. This type of information is, of course, valuable in the planning for better mine safety procedures.

Daily, even hourly, as the Landsats continue to circle the globe, other uses of the pictures taken from space are being found. Soon, a new Landsat will be launched to add to the coverage, and within the next few years a fully operational Landsat system will be working. These satellites give man the means to make better use of the resources of his planet.

11 ENVIRONMENTAL GUARDIANS

Protecting the environment has become one of the most important issues of our time. It is an irony that many of our pollution problems are, in effect, a fallout from our technological progress. The examples are everywhere around us: choking exhaust emissions from cars and trucks; the dumping of wastes into streams and oceans from automated industries and plants; oil spills from modern offshore drilling rigs; and belching smoke-stacks befouling the air at factories turning out the latest lines of products.

Once again, spin-offs from the space program offer potential solutions—especially in the area of setting up a global pollution watch. From space, impartial electronic eyes can see pollution violators and sensitive instruments can monitor earth's ecology.

From an orbital vantage point the entire United States and major portions of the world can be "watched" continuously and automatically. Pollutants can be detected and identified as they move into water. Violations can be pinpointed, and ground crews can follow up on the tips from space by taking samples and determining the causes of the violations.

Satellites can also track air pollution and its distribution patterns over great distances. Concentration levels can be identified, as well as the rate of movement and the rate of dispersion. Such information can be used to tell how far smog or other gases will travel. Pollution alerts, like the storm warnings made possible by weather satellites, could be made in advance.

There are many jobs that can be done economically and efficiently only from space. For instance, monitoring the oceans or the Great Lakes from the ground would require enormous resources and expense—far too costly and time-consuming to be practical. But from orbit, a satellite could provide such information on a large scale, frequently and repetitively, even in remote areas.

Although there are as yet no satellites specifically programmed for environmental tasks, instrumentation on some spacecraft has been used for pollution spotting, monitoring, and tests. The two Landsats are among these.

Experiments aboard Landsat have helped specialists gather information about the quality of water in Florida, Lake Michigan, Lake Erie, Southern California, and in the New York harbor. Ohio State University geologists have used other Landsat photos and data to study long-term effects of strip mining in Ohio, West Virginia, Pennsylvania, and other states. Some mines previously unknown to the investigators appeared in the pictures from space.

In one study in Virginia, Landsat found more than 10,000 smokestacks pouring out particulate emissions. Again, some of these sources had not been known to state officials. Likewise, Landsat has been used to image turbidity variations in Lake Ontario, Lake Champlain, Great Salt Lake, and other bodies of water.

Not only have specific pollutors been pinpointed from space, but imagery has been used in some instances to halt further contamination. One landmark example occurred around Lake Champlain. Dr. A. O. Lind of the University of Vermont found, from an enlargement of a picture taken from space, that a large paper mill on the New York side of the lake was discharging wastes into the water. These wastes were high in sodium and phosphates, which appeared as reddish-brown to a multispectral scanner on the satellite. Dr. Lind also learned that the wastes were being carried across a state line—from the New York side of the lake to the Vermont side. Armed with such foolproof evidence, the State of Vermont took legal action against the paper company and the State of New York, charging that the plant reduced the water quality of the lake below Vermont standards.

There are other random examples of how present-day sensors and techniques—from space—are helping preserve the quality of life on earth. But, to date, such efforts have been limited to relatively small areas. The ultimate goal is continuous coverage of earth from space on a worldwide basis. What is most needed to help accomplish this is a satellite system specifically dedicated to measurements of pollution. Such a satellite is Nimbus G, which is scheduled to be launched sometime in 1978.

One Nimbus G experiment will measure pollutants in the troposphere—the lower atmosphere, in which most of the world's man-made pollution is introduced. The "mixing layer," the part of the troposphere closest to the earth's surface, is where most of the pollutants are trapped—and where we live. This is the domain of local air pollution.

With the launching of Nimbus G and with other sensors and instrumentation on satellites and aircraft, man will have the tools necessary to monitor effectively the world's land, air, and water. No longer will violators be able to pollute without detection, even if they do it surreptitiously under the cover of night. The sources of contaminants will be tracked unerringly and ceaselessly by electronic eyes focused from an all-seeing platform—earth orbit.

12 TECHNOLOGY CLEARINGHOUSE

Part of the Space Act of 1958, under which NASA was created, says that knowledge generated through the national space program be made available to the American people. The law, in fact, states that NASA is required to disseminate its information as widely as possible. Such knowledge and technology, developed with taxpayers' money, should be and is accessible to just about anyone in America who can make use of it.

And in the years since that act was effected, thousands of individuals, private businesses, municipalities, and government agencies have drawn upon NASA's banks of technology to design, develop, and refine new and improved products, processes, systems, and services. Some of these have been described in this book.

To help direct and speed the dissemination of ideas

and information, NASA has established an Office of Technology Utilization. The office uses a variety of ways to do its job. One is through the issuance of a number of publications, which detail new technologies as they evolve, and even point the way, in many instances, to how such technologies can best be used as spin-offs in the commercial marketplace.

Twice a year, for example, NASA publishes a thick book called *Patent Abstracts Bibliography*. This lists hundreds of NASA-owned inventions which are available for licensing. Continuing series of "Tech Briefs" papers also are printed and distributed widely—more than 50 million to date—covering subjects ranging from the latest developments in flame-retardant fabrics to new techniques in industrial welding. All of these are based on research initially done for specific space projects.

One Tech Briefs bulletin about NASA's research on a practical solar heating and cooling system for housing, published in 1974, was requested by more than 4,000 individuals and firms.

Additionally, experts write, edit, and publish scientific and technical aerospace reports and journals, and these, too, are made available to the public. To further ease the flow of information into popular uses, NASA, in 1963, set up a network of six Industrial Application Centers at universities spread geographically across the United States. These centers today tie into what has become the largest storehouse of scientific and engineering data in the world. In fact, these centers have com-

Information on call: one of the nation's largest computerized software data libraries of engineering analyses is maintained by NASA at the Computer Software Management and Information Center (COSMIC) at the University of Georgia.

puterized access to more than seven million documents. And this list of documents is growing at the rate of about fifty thousand new documents every month.

All of this vast information is available to any individual or company, regardless of how large or small it is, for a modest service charge. Say a company is thinking about producing a new, pocket-sized electronic calculator, or a new line of digital watches. Before making the decision to manufacture such products, the company's leaders would want to know the latest developments in microelectronic circuits and chips and other components that would go into these calculators or watches. They would ask questions at one of the centers. Staff members there would conduct a computer search, which would reveal everything in the data storehouse pertaining to the subject, and this, in the form of documents, research reports, or whatever, would be given to the company.

Millions of technical reports can be reviewed in a matter of hours. Such a system saves businessmen precious time and millions of dollars that they might otherwise have had to spend to do their own research. Thousands of companies use these center services annually.

Additionally, a highly skilled staff of specialists in such fields as chemistry, physics, electronics, and pharmaceuticals is available at each center to assist people seeking information, and to help them solve particular problems.

"Every information search we perform for someone or some company helps NASA fulfill its charter and may lead to a profit for the company and a better product or service for the American people," says Jeffrey T. Hamilton, Director of NASA's Technology Utilization Office.

Anyone wanting information on just about any technical subject imaginable can find out how and where to get it by writing this Office, care of NASA Headquarters, Washington, D.C., 20546.

13 A BETTER TOMORROW

Most of the space spin-off benefits discussed in this book are either in everyday use now, or are in advanced stages of testing and experimentation and will soon be available. But what of the longer-term future? What can be expected through the rest of this century and beyond? What can you expect in your lifetime?

The fallout of space-developed technology will become more and more important to us in almost every phase of our lives. It will result in better, safer, more reliable, and longer lasting products—everything from batteries and automobile tires, to newer, cleaner, and less expensive means of heating and cooling our homes.

We will be able to communicate, with both voice and picture, to anyone, anywhere in the world, by way of satellite. Farmers will be able to produce greater amounts of needed food more efficiently. Improved transportation methods, leading to the era of rocket travel, will speed us to distant destinations safely and swiftly. Through ever-watchful electronic eyes in space, we will find new sources of raw materials essential to keep our mechanized, automated society moving forward. We will learn more about the weather; how to forecast it more accurately for longer periods of time, and, eventually, how to control it for the benefit of all people on earth. We will learn more about the precious planet on which we live and how to protect its environment from pollution that endangers future generations.

Many major resources, still to be exploited through the application of space processes and systems, are contained in the world's oceans. They represent an important source of food and minerals. The circulation within the seas has a decisive influence on this food source through the transporting of nutrients and oxygen, control of temperature, and salinity, or salt content.

Yet our present overall knowledge of the oceans is relatively primitive. One of the main reasons for this is the sheer size of the oceans and other bodies of water that cover more than 70 percent of the earth's surface. For example, collection of global oceanographic data on a regular—perhaps weekly or daily—basis using ships or aircraft simply is not practical.

Satellites from space, however, are uniquely suited for continuous observation and monitoring, and NASA is now working on such a spacecraft. It is called Seasatellite, or Seasat for sho.t. As the first truly oceanographic satellite, it will be capable of providing constant, worldwide, timely information on global ocean dynamics and other physical properties of importance.

How will this data be used? Predictions of wave heights and wind fields will be useful in planning ship routing, ship design, storm damage avoidance, coastal disaster warning, and offshore power-plant siting. Maps of current patterns and temperatures will be helpful in ship routing, fishing, pollution dispersion, and iceberg hazard avoidance. Charts of ice fields will be used for navigation and weather prediction. Seasats in orbit above earth will keep watch on the oceans much as the Landsat satellites are surveying land and sea today.

In the years to come, we will build new space transportation systems that greatly reduce the costs of getting to and from earth orbit, and to and from other planets. We will explore other celestial bodies in our solar system and mine them for valuable minerals and other resources needed on earth. We will place temporary space stations in orbit, and follow these with permanent stations, manned by scientists, engineers, meteorologists, geologists, oceanographers, and others.

New materials and products, including vaccines, will be produced on such stations free from the restraints of earth's gravity. This will lead to the day of orbiting factories and hospitals. We will colonize the moon, Mars, and other planets, and learn from their history and development how to better manage and protect earth. Beyond this, electronic voyagers, followed by manned spacecraft, will be dispatched to the far reaches of our solar system, and then farther into space to seek new worlds, new opportunities, and, quite possibly, to find other life.

It would be easy to scoff at such fantastic dreams if we did not look at the record of accomplishment to date. The Space Age really began only a few years ago—in 1957, when the first small satellite was launched into earth orbit. Yet less than twelve years later, man explored the moon, and by the mid-1970s, instruments had been landed on Mars and Venus, and spacecraft have flown past the faces of Mercury, Jupiter, and Saturn.

Thus it will not be a lack of technology that will limit our progress. More likely, it will be money that determines how far and how fast our space program advances. It is expensive to launch rockets and send out spacecraft. Most of this money comes from American taxpayers and is budgeted by the federal government. There are many essential needs for this money here on earth. Only a very small percentage of it can be allotted each year to fund NASA's programs.

So NASA must do what it can to make the most of these limited funds. Much emphasis now is being placed directly on this problem. The most important result is development of an entirely new and far less expensive means of transportation to and from earth orbit. This system is called the Space Shuttle.

Until now, NASA has launched satellites into orbit, or sent spacecraft to the moon and to other destinations, on a one-time-use basis. That is, once the rockets launch the

Orbital transportation system of the future: the versatile Space Shuttle, scheduled for missions in the 1980s, will be able to fly to earth orbit and return to land much like a modern jet airliner. Because it can be reused again and again, it will greatly reduce the cost of space flight, and will be capable of performing a variety of jobs. In this artist's concept, the Shuttle releases a satellite from its giant cargo bay.

spacecraft, they are discarded. And once the satellite has performed its job in space, it is never used again. Each rocket and each spacecraft costs millions of dollars.

For years, NASA has been working on a new system where the same basic launch vehicle could be used over and over again, and satellites in orbit could be serviced and repaired to extend their useful lifetimes.

The shuttle will do this. NASA expects to have this versatile new vehicle in use early in the 1980s. At launch, the shuttle's orbiter will be mated to a large propellant tank and to two solid-fueled rocket boosters. The orbiter is the key to the system. It is about the size of a DC-9 jet liner and looks much like a delta-winged airplane. A huge cargo compartment in the orbiter will be used to ferry a wide assortment of "payloads" to and from earth orbit. This compartment is about 15 feet in diameter and 60 feet long. It is large enough to haul 65,000 pounds into an orbit 115 miles above earth.

At lift-off, the twin rocket motors will ignite simultaneously, and burn in concert with engines on the orbiter. Once the shuttle is well on its way toward space, the solid rockets will drop off and return to earth by parachute, where they will be recovered, fixed up, and used again.

The orbiter, with the large propellant tank still strapped to it, will continue on until earth orbit is reached. The tank will then be jettisoned, and drop off into the ocean. The orbiter will remain in space with its crew of four or more astronauts and its cargo for an average of about seven days, although this can be stretched to a month if required.

From its station in space, the orbiter will be able to perform a number of valuable functions. A new weather communications satellite could be brought up with it, then released from the cargo compartment after earth orbit has been gained. A satellite already in orbit that has malfunctioned could be serviced on the spot by experts. Repairs could be made and it could be placed back in service—a feature never before possible. Or, if more extensive work is necessary, a satellite could be stored in the orbiter and brought back to earth.

In many ways, the orbiter can be looked upon as a sort of space truck. It may be used to haul large and heavy components for a space station into orbit. Or another whole rocket system could be carried above earth by the orbiter, be reassembled there, and then be used to launch astronauts to distant planets.

Whatever the assignment, when it has been completed in space, the orbiter will be flown back to earth. It will maneuver as a spacecraft until it reaches a point about 400,000 miles above earth. From this altitude on, aerodynamic surfaces and controls will take over, and the orbiter becomes in effect an airplane, which will "glide" through a reentry into the earth's atmosphere and land on a specially built runway at the Kennedy Space Center in Florida.

The orbiter then will be refurbished and readied for another mission, possibly within two weeks. Each orbiter is being designed and built to make a high number of such earth-to-orbit and return flights—possibly fifty or more.

The shuttle will become our all-purpose space transportation system of the future. In the long run, it will be much more versatile—and far less expensive—than any present system, primarily because it will be able to do so many more things, and because it will be reusable.

Beyond the shuttle, in time, will come space stations. Man already has proved he can remain in earth orbit for relatively long periods of time without harm, and can perform many useful purposes. The Skylab series of missions in the early 1970s verified this when three separate teams of astronauts each were launched to an experimental station above earth and stayed for periods of up to nearly two months.

One of the most intriguing series of Skylab tests in

Skylab series of flights in the early 1970s proved that man could live and work comfortably in the weightless environment of space for relatively long periods of time—up to two months. Here astronaut Owen Garriott operates at a work station aboard Skylab 3.

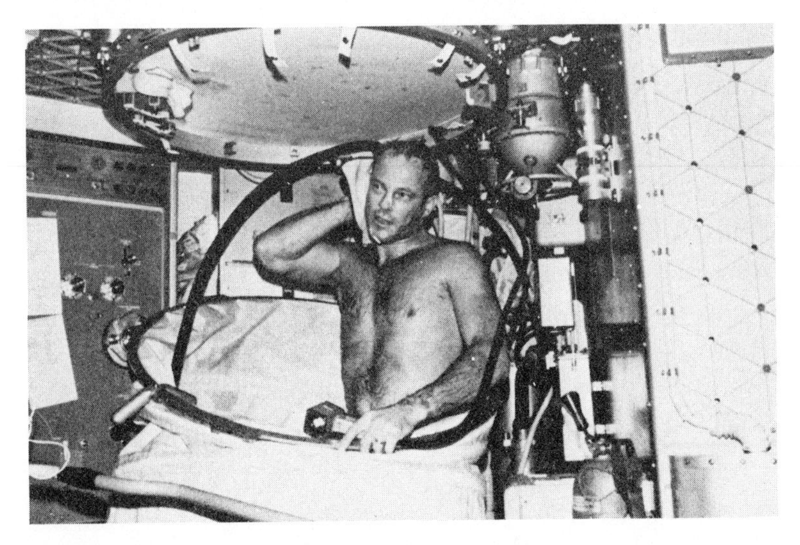

Space shower: astronaut Jack Lousma bathes in the weightless environment of space in earth orbit aboard Skylab 3. Lousma and two other astronauts performed a number of medical, scientific and technological experiments on Skylab, proving that man will be able to work effectively in space on the long-term missions of the future.

orbit was aimed at studying the effects of weightlessness, or gravity-free environment, on certain manufacturing processes. In one such experiment, astronaut Jack Lousma fired up an electrical furnace to form new metal alloys that will not blend on earth because of gravity. It is believed that perfect spheres, such as ball bearings, can be cast in a weightless environment; pure crystals can be grown; and foamed, high-strength metals can be made. All of these are not possible to produce on earth at the present time.

Experts also say that the absence of gravity in space will one day allow the production of many medical drugs at less cost and greater purity. The separation of viruses or bacteria for the making of disease-fighting vaccines is a difficult, multistep process on earth. Such separation in a state of weightlessness, however, can be done faster, and more simply, efficiently, and economically.

These representative samplings of known improved manufacturing processes and products and biological processing techniques are just the tip of the iceberg. From space stations permanently orbiting earth would come scores of other discoveries and applications leading to far-reaching benefits for people on earth.

The day of orbiting factories, power stations, and hospitals will come, once the technologies have been mastered and the costs have been made competitive. And in so doing, earth will further be protected from the threat of ever-increasing industrialization, and the inevitable pollution this causes.

Space stations of the future will provide magnificent

Among the many exciting space projects of the future now under study is a solar-satellite power station, as depicted in this artist's concept. Such a station would convert the sun's radiation into electricity by large solar arrays. This energy would be transformed to microwaves and transmitted to earth where a receiving station would convert the incoming microwaves into DC electricity at a very high efficiency. This satellite concept is one of several methods being explored as ways to generate large amounts (megawatts) of electric power for uses on earth.

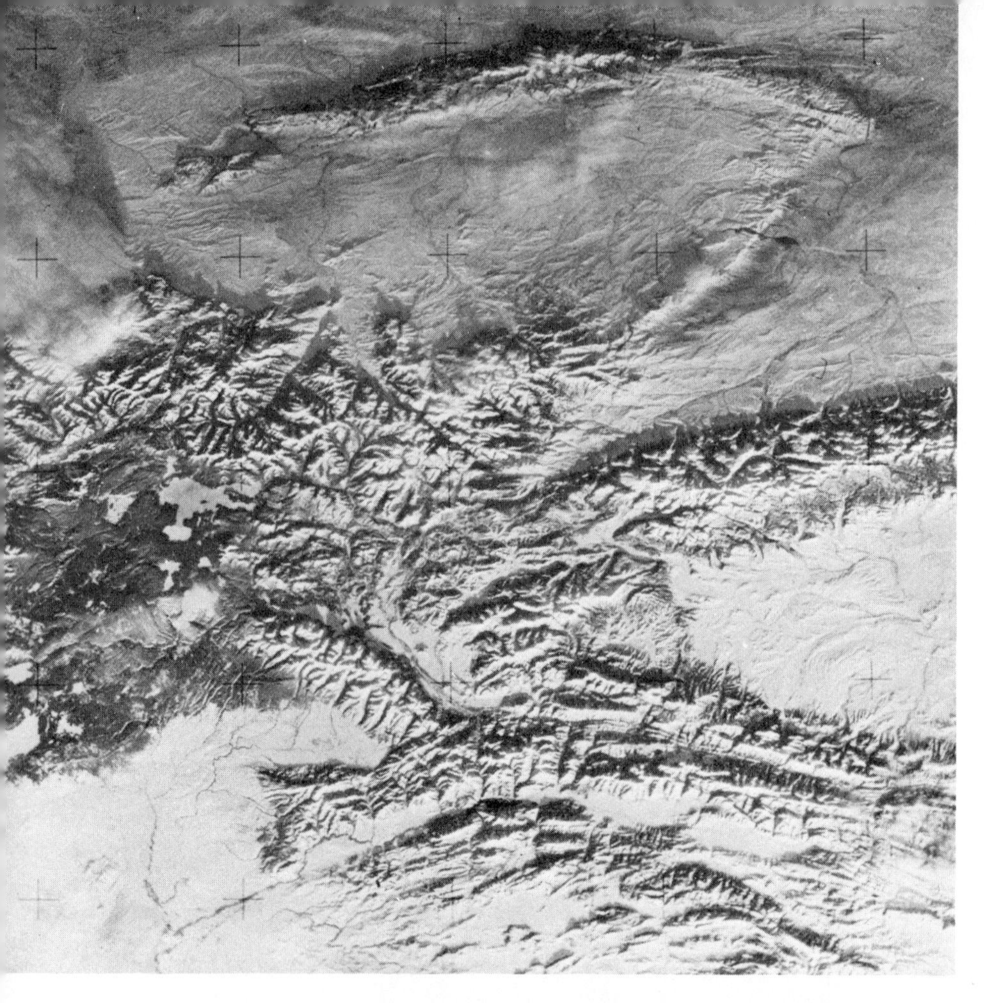

Snow-covered northwest corner of Wyoming is seen in this photo taken from earth orbit aboard Skylab 4. In the future, scientists and experts on space flights will be able to observe earth from such a spectacular vantage point, and report their findings immediately to others on earth.

scientific platforms and laboratories from which scientists in many fields can observe our planet. Consider, for example, what improvements in weather forecasting could come from having a professional meteorologist keeping a watch on weather phenomena from a station in space.

Consider what it would be like to have foresters, hydrologists, oceanographers, agricultural experts, geologists, and others making constant, up-to-the-second observations, and interpreting and analyzing the raw information on the spot. And all of what they see and forecast from the vantage point in space could be relayed instantly to the appropriate outlets on earth. Forest fires could be spotted almost immediately; great schools of fish could be sighted and commercial fleets directed to them; severe storms could be watched throughout their development and their exact course tracked, with adequate warnings flashed at each turn.

Next could come a space colony. One proposal features a wheel-shaped design that would house up to ten thousand people, along with shops, schools, light industry, and a self-contained agriculture system for producing food. The principal industries would be the manufacture of more such colonies and the construction of solar energy collectors that would be placed in orbit near the earth to beam down cheap energy. Solar energy would power the space colony, too. Heavy industry would be conducted outside the living area to make use of the weightlessness and vacuum of space.

After completion of the first colony, larger ones could be built, some orbiting farther from earth. The material of the asteroids, between Mars and Jupiter, would be sufficient for the construction of colonies with a total land area thousands of times greater than all of earth's continents. On the moon, Mars, Mercury, and the asteroids are treasure houses of minerals and all the other elements that human minds and tools will need to sustain and advance life for centuries to come.

Beyond that lie the stars. Why bother? Although it is unlikely that we will find intelligent life on the other planets of our sun, many scientists believe it is likely that we would find life among the stars of the galaxy. The most feasible way to initiate contact with other life in space would be to listen for radio signals. It will be years, perhaps decades, before we could send instrumented spacecraft on such distant voyages.

Efforts are already under way to tune in signals emitted from elsewhere in the universe. Additionally, NASA has plans to build a multichannel spectral analyzer under a program known as SETI (Search for Extraterrestrial Intelligence). The device will be used with various radio telescopes to sweep across space toward stars, and any planets that might surround them, in hopes of making contact in a radio frequency band.

The odds may be overwhelmingly strong against making such contact within our lifetime. But without trying, there would be no odds at all.

Meanwhile, as we send spacecraft to explore the surface of neighboring bodies in our solar system, and as we prepare to listen for signs of other intelligent life far out in the universe, the spin-off benefits from the nation's space programs of the past twenty years continue to improve our life on planet earth.

FOR FURTHER READING

Material for this book was obtained from a variety of sources, including interviews with NASA officials and executives of aerospace companies; NASA Congressional testimony; NASA and aerospace company documents, brochures, booklets, fact sheets, annual reports, and other literature.

For further reading, the following is a list of key source materials and how and where to obtain them.

These publications may be obtained by writing the Superintendent of Documents, U.S. Government Printing Office, Washington, D.C. 20402:

"Aeronautics: Space in the Seventies." Number 3300-0409. 75¢.

"Energy-Related Research and Development." Special report-prepared for the Committee on Aeronautical and Space Sciences, U.S. Senate. April 1975. $1.55.

"NASA and Energy." 1974. Number 3300-00567. 35¢.

"New Horizons." 1975. Number 033-000-00631-4. $2.00.

"Technology Utilization Program Report." NASA, 1975. Number 3300-00594. $2.10.

"Toward a Better Tomorrow with Aeronautics and Space Technology." Special report prepared at the request of the Committee on Aeronautics and Space Sciences, U.S. Senate. 1973. $1.55.

The following information can be obtained, free of charge, from the NASA Public Information Office, 400 Maryland Ave., S.W., Washington, D.C. 20546:

"Medical Benefits from Space Research." 1975.

NASA Fact Sheets: "Solar Energy Applications" and "Solar Energy for Heating and Cooling of Buildings." 1975.

"Photography from Space to Help Solve Problems on Earth." 1974.

"Space Applications." 1975.

"Space and Man's Environment." 1974.

"Spaceship Earth: A Look Ahead to a Better Life." 1974.

"Spinoff 1976: A Bicentennial Report." 1976.

The following information can be obtained from the

NASA Education Office, 400 Maryland Ave., S.W., Washington, D.C. 20546:

"NASA Facts: Aeronautics." 1974.
"NASA Facts: Space Benefits." 1974.

The following information can be obtained from the NASA Technology Utilization Office, 400 Maryland Ave., S.W., Washington, D.C. 20546:

"Applications of Aerospace Technology in the Public Interest: Pollution Measurement." 1975.
"Space Benefits Briefing Notebook." 1974.
"Space Benefits: The Secondary Application of Aerospace Technology in Other Sectors of the Economy." 1976.

The following information may be obtained by writing the Communications Satellite Corporation, Public Information Office, 950 L'Enfant Plaza, S.W., Washington, D.C. 20024:

"Communications Satellite Corporation Report to the President and the Congress." 1975 and 1976.
"Comsat General Marifacts," magazine article, October 1975 issue.
"Comsat's Pocket Guide to the Global Satellite System." 1975.

The following information may be obtained by writing to these companies:

"Landsat: NASA Earth Resources Satellite." General Electric Company, Space Division, Valley Forge Space Center, P.O. Box 8555, Philadelphia, Pa. 19101. 1975.

Annual reports, 1974 and 1975, and magazine article, "Closing the Gap in U.S. Space Communications," Western Union Communicator Magazine, Summer 1975. Western Union Corporation, 85 McKee Drive, Mahwah, N.J. 07430.

Also:

Ordway, F. I., III; Adams, C. C.; and Sharpe, M. R., Jr. *Dividends from Space.* Thomas Y. Crowell, New York. 1971.

Taylor, L. B., Jr. *For All Mankind: America's Space Programs of the Seventies and Beyond.* E. P. Dutton, New York. 1974.

INDEX

aeronautical research, 49–52
environmental protection in, 51
fuel conservation in, 49–50
aerosats, 48
agricultural surveys, 90–92
airports:
fog hazards at, 29
"grooving" runways of, 26
traffic congestion at, 51
air-traffic control, 29–30, 38, 48, 51
air travel, rocket-powered, 48, 107
Aldrin, Edwin, 1
aluminum foils, 32
Ames Research Center, 4, 25
amputees, 19
Anik satellites, 66
Apollo missions, 1, 11–12, 23, 27, 39, 73
APT (Automatic Picture Transmission), 77
Armstrong, Neil, 1
artificial limbs, 14, 19
asteroids, 117
ATS (Applications Technology Satellite), 5, 68–72
automotive industry, 46

baggage carts, 33
bank credit systems, 38
banking machines, 39
batteries, rechargeable, 16, 33
Bay Area Rapid Transit System (BART), 44–45
bedridden patients, 12, 20
biotelemetry, 15
blackouts, 40
blankets, 34–35
bleeding, internal, 4–5

blind persons, electronic devices for, 14, 20–21
braces, orthopedic, 20
bridges, detecting damage to, 46
building industry, 42
bullets, identification of, 43
burn victims, 11–12

cables, submarine, 63
calculators, electronic, 41, 104–105
cameras, electrostatic, 14
cerebral palsy, 15
check processing, 39
clocks, digital, 33
clothing:
  for firemen, 23
  for sportspersons, 34–35, 36
Coast Guard, U.S., 25–26, 28
Communications Satellite Act, U.S. (1962), 65
Communications Satellite Corporation (Comsat), 65
communications satellites, 32, 42, 61–73
  capabilities of, 62, 64–66, 106
  commercial development of, 65–68
  domestic, 66–68
  earth stations and, 64–65
  education via, 68–72, 73
  medical data sent by, 5–6, 71
  in synchronous orbits, 63–64

composite materials, 41, 50
compressed air breathing systems, 23–25
computers:
  in law enforcement, 42
  software for, 37–40, 104
  in transportation, 44, 45, 50
Computer Software Management and Information Center (COSMIC), 104
Comstar, 66
constant-temperature shields, 11–12
consumer items, 31–35, 36
cooking pins, thermal, 32
credit authorization, 38, 39
crime prevention and detection, 42–43

dams, inventory of, 96
data processing, 37–40
department stores, computers used by, 39–40
design engineering, 46
dry immersion beds, 12

Early Bird I, 65
earthquake hazard maps, 96–97
Echo, 32
echo-cardioscopes, 17
education:
  by satellite, 68–73
  in space achievements, 72–73
electrical power sources, 55–60, 115

electrical wiring, 34
electric power dispatch terminals, 40
electrocardiograms (EKG), 6–7
electroencephalographic (EEG) tracings, 11
electromagnetic radiation, 86
electronic calculators, 41, 104–105
electronics (*see also* computers; microelectronics), 6–11, 14, 20–21, 45, 47
electronic strain gauges, 41
electrooptics techniques, 42
energy, 53–60
  geothermal sources of, 54–55
  shortages of, 2, 53–54, 60
  solar, 42, 55–60, 115
  wind, 42, 55, 56, 57
"energy farms," 59–60
Energy Research and Development Administration, 54–55
environmental protection, 2, 51, 98–101, 107
  by Landsat, 99–100
  in Nimbus G experiment, 101
Explorer I, 1
eye tumors, removal of, 21

facsimile transmission, 62, 68
Federal Aviation Administration, 29
Federal Bureau of Investigation (FBI), 42
Federal Highway Administration, 45–46
fire-fighting modules, 25–26
firemen, apparel and equipment of, 23–25
fire-resistant materials, 25
fire safety for volatile fuels, 25
flashlight switches, 33
flood monitoring, 95–96
food processing and preparation, 31–32
food supply, 90, 106–107
forest fires, 93, 117
forest surveys, 93
freeze-dried foods, 31–32
fuel conservation, 49–50

Garriott, Owen, 113
geological surveys and studies, 84, 87, 88, 96–97, 99–100
golf carts, 33
gravitation, 62
"greenhouse effect," 58

Hamilton, Jeffrey T., 105
handicapped, 16–21
hands, artificial, 19
hearing defects, 11, 14
heart, sonar monitoring of, 13, 17
heart attacks, 6–7
heart pacemakers, 14, 16, 33
heat shields, 19, 32

helmets, astronauts':
antifogging compound for, 27
medical use of, 11
helmet visors, 27
highway safety, 26–27
crash barriers for, 27
visibility and, 27
home, space technology in, 31–35
home construction, 34, 42
hurricanes, 74–76
hydroplaning, 26

illumination, crime-deterrent, 43
Industrial Application Centers, 103–104
industrial process control, 38
industry, 37–43
computer technology in, 37–40
microelectronics in, 40–41
NASA and, 102–105
sample spinoffs in, 41
in weightless environment, 114
infants, breathing problems of, 7–9
insurance companies, 38
Intelsat (International Telecommunications Satellite Consortium), 65
Intelsat IV-A, 65–66
ITOS (Improved Tiros Operational Satellites), 77–78

jackets, 35, 36
Jet Propulsion Laboratory, 54
Johnson Space Center, 25, 43
Jupiter, 1, 109

Kennedy Space Center, 18, 22–23, 112

Landsat satellites, 54, 82–97, 108
agricultural surveys by, 90–91
capabilities of, 82–85
earthquake zones observed by, 96–97
ecosystems surveyed by, 93–94
habitat distribution surveyed by, 92
operation of, 85–86
pest control aided by, 91–92
pollution control by, 99–100
resources sought by, 54, 84, 88–90
water-resources and flood monitoring by, 94–96
land-use planning and management, 96
Langley Research Center, 59
Composite Materials Laboratory of, 19–20
"Technology House" at, 42
laser surgery, 21
law enforcement, 42–43
leg braces, 20

Lewis Research Center, 55
life, extraterrestrial, 118
life-raft canopies, 34, 35
life rafts, radar-reflective, 28
Lind, Dr. A. O., 100
Lousma, Jack, 113, 114

Mailgrams via satellite, 62, 68
Marisat (maritime satellite system), 68, 69
Mars, 2, 73, 108, 109, 117
Marshall Space Flight Center, 19
medical care, 4–21
  for bedridden, 12–13
  in cardiac cases, 6–7, 13, 16, 33
  communications in, 5–6
  for handicapped, 16–21
  for internal bleeding, 4–5
  for paralysis victims, 17–19
  patient-watching duties in, 7–8
  research and testing in, 13–17
  robot technology in, 16–19
medical drugs, 114
Mercury, 1, 109, 117
microelectronics, 16, 40–41, 44, 105
micrometeorites, 13
microwave radio stations, 63
mineral sources, 84, 88–90, 107, 108, 117
mining, 97, 99–100
mirrors, space, 43
momentum transducers, 13
moon:
  colonization of, 108
  earth seen from, 73
muscle accelerometers, 13–14
muscle-oriented devices, 18–19
Mylar, 34–35, 36

NASA (National Aeronautics and Space Administration), 3, 32, 33, 41, 45–46, 97
  aeronautical research of, 49–52
  education and, 68–73
  energy improvement programs of, 54–59
  finances of, 109
  future of, 107–118
  medical care and, 4, 5, 6, 7, 13, 16–21
  safety equipment of, 23–30
  as technology clearinghouse, 102–105
  weather satellites of, 76–81
NASTRAN, 40, 46
National Advisory Committee for Aeronautics (NACA), 49
National Crime Information Center (NCIC), 42
navigation, 46–48, 68, 69, 108
neurological ailments, 14
Newton, Isaac, 62
nickel-cadmium battery cells, 16, 33

Nimbus satellites, 77, 101
nylon, heat-reflective, 34

oceanographic data, 107–108, 117
oil, *see* petroleum sources
oil spills, 98
oxygen analyzers, 11

pacemakers, 14, 16, 33
paints, 33, 41
paralysis, 17–19
Parkinson's disease, 14
Patent Abstracts Bibliography, 103
pelvic braces, 20
pest control, 91–92
petroleum sources, 54, 84, 88
photographs:
  by Landsat, 85, 86, 87, 92, 94, 96
  by weather satellites, 75, 76, 77, 80
plastic foam, 12
pollution, 2, 53, 98–101, 107
polyester film, transparent, 32
pressure suits, 5, 8, 9
proximity suits, 23
publications, 103–105

quadriplegics, 17
quartz crystal oscillators, 33
"Quietports," 51

radios, pocket-sized, 29
rail transportation, 44–45, 96
Randomec, 46
rangelands, search for, 91
raw materials, 54, 84, 88–90, 107, 108, 117
remote sensing, 86, 92
respiratory ailments, 7–9
respirometers, helmets as, 10, 11
retail sales systems, computerized, 38, 39–40
robot technology, 16–19
rocket nozzle liners, 19
rocket-powered air travel, 48, 107

SAFER (Systematic Aid to Flow on Existing Roadways), 45
safety, 22–30
  in aviation, 29–30
  in fire-fighting and prevention, 23–26
  on highways, 26
  in mines, 97
  on waterways, 27–28
satellites (*see also* communications satellities; Landsat satellites; weather satellites):
  in air-traffic control, 29–30, 48
  energy sources sought by, 54, 84, 88
  launching and orbiting of, 1, 62–63
  in law enforcement, 42
  in navigation, 46–48, 68, 69, 108

satellites (cont.)
  oceanographic data from, 107–108
  pollution monitored by, 98–101
  as solar power stations, 60, 115
  Space Shuttle and, 111
Saturn, 109
sealants, 33
Seasatellite (Seasat), 107–108
seats, spacecraft, 12
sensors, medical monitoring with, 7–9
SETI (Search for Extraterrestrial Intelligence), 118
ships, "optimum-time-routing" of, 48
sight switches, 17–18
ski parkas, 34, 35
Skylab, 112–114, 116
sleeping bags, 34, 35
social problems, space programs and, 2
solar energy, 55–60, 103, 117
  collector systems for, 42, 58–59
  satellites for, 60
  solar cell arrays for, 59–60, 115
  timetable for, 60
sonar machines, 13, 17
Space Act (1958), 49, 102
space colonies, 117–118
space programs:
  educational role of, 72–73
  expense of, 2–3, 108
space programs (cont.)
  future spin-offs from, 106–118
Space Shuttle, 109–112
space stations, 108, 111, 112–117
  earth observed from, 116–117
  manufacturing in, 112–114
space suits, 5, 8, 9
splints, 12
sportspersons, fabrics for, 34–35, 36
Sputnik I, 1
steel alloy, precipitation-hardened, 41
Stormsat, 80–82
strip mining, long-term effects of, 100
surgical masks, helmets as, 11
survival kits, astronauts', 28
Synchronous Meteorological Satellite (SMS) program, 78–80
synchronous satellite orbits, 63–64

taffeta, heat-reflective, 34
"Tech Briefs," 103
Technology Utilization Office (NASA), 103, 105
Teflon, 32
teleoperator and robot technology, 16–19
telephone, transoceanic, 65, 66

television:
  transoceanic, 61, 62, 64, 65, 66, 71–72
  by weather satellite, 77, 78
thermos jugs, 32
tires:
  "heat pictures" of, 27
  studless, 46, 47
Tiros (Television and Infra Red Observation Satellites), 77
tongue-pressure switch, 18
toothpaste, 12–13
tracheotomies, 7
traffic control systems, 45
TRAIN II, 45
transducer-transmitters, 14
transmitters, medical monitoring with, 7–9
transportation systems, 44–52, 96
  air, 29–30, 38, 48–52, 107
  highway, 26–27, 46
  rail, 44–45
  ship, 46–48, 68, 69, 108
  space, 108–112
troposphere, pollutants in, 101
Trueblood, Dr. H. Wood, 4

ulcers, 12
ultraviolet radiation protection, 33
underwear, 34

Venus, 1, 109
vibration sensors, 41
V/STOL (Vertical and Short Takeoff and Landing) aircraft, 51

watches, digital, 33, 105
waterfowl habitats, 92
water pollution, 98, 99, 100
water-resources monitoring and management, 94–95
weather forecasting on space stations, 116
weather satellites, 74–81, 111
  agriculture and, 81, 90
  development of, 76–81
  long-term forecasts by, 81, 107
  short-term forecasts by, 80–81
weightlessness, 12, 13, 113, 114
Westar satellites, 66–68
wildlife surveys, 92, 94
windmills, electrical power from, 55, 56, 57
windpipe obstructions, 7
windshields, automobile, 27
windshields, spacecraft, 11–12
wind-tunnel experiments, 33
wiring, electrical, 34

X rays, 14

ABOUT THE AUTHOR

L. B. Taylor, Jr., has worked in and around the space industry for many years, at Cape Canaveral and the Kennedy Space Center during all the major manned and unmanned launches. He is also the author or coauthor of three previous books on aerospace subjects, and one on buried treasure. In addition, he has had more than 200 articles published in major national magazines.

An avid traveler, Mr. Taylor makes his home in Williamsburg, Virginia, with his wife and three children.